PROACTIVE HUMAN–ROBOT COLLABORATION TOWARD HUMAN-CENTRIC SMART MANUFACTURING

PROACTIVE HUMAN–ROBOT COLLABORATION TOWARD HUMAN-CENTRIC SMART MANUFACTURING

SHUFEI LI
Department of Industrial and Systems Engineering
Hong Kong Polytechnic University
Hung Hom, Kowloon, Hong Kong

PAI ZHENG
Department of Industrial and Systems Engineering
Hong Kong Polytechnic University
Hung Hom, Kowloon, Hong Kong

LIHUI WANG
Department of Production Engineering
KTH Royal Institute of Technology
Stockholm, Sweden

ELSEVIER

Elsevier
Radarweg 29, PO Box 211, 1000 AE Amsterdam, Netherlands
125 London Wall, London EC2Y 5AS, United Kingdom
50 Hampshire Street, 5th Floor, Cambridge, MA 02139, United States

Notices

Knowledge and best practice in this field are constantly changing. As new research and experience broaden our understanding, changes in research methods, professional practices, or medical treatment may become necessary.

Practitioners and researchers must always rely on their own experience and knowledge in evaluating and using any information, methods, compounds, or experiments described herein. In using such information or methods they should be mindful of their own safety and the safety of others, including parties for whom they have a professional responsibility.

To the fullest extent of the law, neither the Publisher nor the authors, contributors, or editors, assume any liability for any injury and/or damage to persons or property as a matter of products liability, negligence or otherwise, or from any use or operation of any methods, products, instructions, or ideas contained in the material herein.

ISBN: 978-0-443-13943-7

For information on all Elsevier publications
visit our website at https://www.elsevier.com/books-and-journals

Publisher: Matthew Deans
Acquisitions Editor: Chiara Giglio
Editorial Project Manager: Tessa Kathryn
Production Project Manager: Sujithkumar Chandran
Cover Designer: Vicky Pearson Esser

Typeset by VTeX

Working together
to grow libraries in
developing countries

www.elsevier.com • www.bookaid.org

Contents

10. Conclusions and future perspectives   **265**

List of figures

List of tables

Preface

Proactive Human–Robot Collaboration (Proactive HRC) fosters a mutually beneficial partnership between human and robotic agents, leveraging their complementary capabilities, in line with the human-centric paradigm of smart manufacturing. Employing advanced cognitive computing, knowledge graphs, large language models, and robot learning technology, Proactive HRC mirrors human intellectual capabilities in intricate manufacturing scenarios. Unlike traditional HRC, which relies on predefined procedures, Proactive HRC dynamically senses its surroundings, comprehensively acquires manufacturing knowledge, proactively plans task sequences, and optimally executes tasks. It adheres to the flexibility design principle, enabling it to adapt to changing work settings and task requirements. Characterized by mutual-cognitive, predictable, and self-organizing intelligence, the robotic component within the Proactive HRC system functions as an extension of the human body, actively coordinating with humans and performing dexterous tasks.

Enabled by intelligent robot control and a human assistant system, Proactive HRC facilitates the collaboration of multiple humans in various roles and robotic agents with diverse workloads on complex manufacturing tasks in close proximity, fostering bi-directional trust. This collaborative approach takes into consideration each other's operational needs, resource requirements, and complementary capabilities. The intelligent capabilities of Proactive HRC find broad applications in disassembly, immersive drilling, and advanced welding tasks, capitalizing on the agility, problem-solving abilities, and creativity of humans, combined with the speed, accuracy, and precision advantages of robots.

The integration of Proactive HRC into production processes enhances the well-being and job satisfaction of human workers, creating more comfortable and safer working conditions while boosting overall productivity. Both humans and robots within the Proactive HRC actively explore diverse solutions and manipulation approaches for completing various manufacturing tasks, effectively addressing challenges within current production systems.

In light of these considerations, this pioneering book, the first of its kind in the field of Proactive HRC, sets out to elucidate its core principles and delve into state-of-the-art enabling technologies that underpin proactive collaboration between multiple humans and robots in today's smart manufacturing environments. The ultimate goal is to advance human-centric, sustainable, and resilient prospects within this domain. A significant portion of the book's content is rooted in several research projects conducted in Hong Kong and Sweden. It offers a comprehensive compilation of innovative findings and practical examples aimed at highlighting the unique attributes of Proactive HRC and facilitating its widespread implementation in industrial contexts.

Furthermore, Chapter 8, which provides practical operational mode examples of Proactive HRC, draws from the research work of Mr. Jianzhuang Zhao and Dr. Edoardo Lamon at the Istituto Italiano di Tecnologia in Italy, while the disassembly case of an aging electric vehicle battery in Chapter 9 is the result of Dr. Rong Zhang's efforts at Donghua University in China. This book is intended to serve as a valuable reference for university students taking advanced courses, particularly in the fields of industrial engineering and business management. The authors also aspire to capture the interest of an ever-expanding community of researchers and practitioners in this promising field. They welcome open discussions, in-depth research, and development efforts to further enhance the practical implementation of Proactive HRC in the era of human-centric smart manufacturing.

For additional information and videos please visit the link https://www.raids.group/proactive-human-robot-collaboration/.

Acknowledgments

The authors express their sincere gratitude to several key members of the research team at The Hong Kong Polytechnic University. Specifically, Mr. Junming Fan and Mr. Chengxi Li played essential roles in developing the methodologies for Proactive HRC, while Mr. Hongpeng Chen offered valuable solutions in the context of Proactive human–robot collaborative drilling in Chapter 9. Additionally, the authors are grateful to Ms. Tian Wang and Ms. Wenhang Dong for their insights into the future perspectives of Proactive HRC, as discussed in Chapter 10.

Meanwhile, the authors are also grateful for the funding support from the National Natural Science Foundation of China (No. 52005424), Research Grants Council of the Hong Kong Special Administrative Region (Project No. PolyU 15210222), Endowed Young Scholar in Smart Robotics (Project No. 1-84CA), Research Committee of The Hong Kong Polytechnic University under Research Student Attachment Programme 2021/22, Collaborative Departmental General Research Fund (G-UAMS) from the Hong Kong Polytechnic University, Hong Kong Special Administrative Region, the open program at the State Key Laboratory of Intelligent Manufacturing Equipment and Technology (No. IMETKF2024010), China, and the EU H2020 ODIN project (Grant Agreement: 101017141).

Finally, we extend our gratitude to the colleagues at Elsevier, specifically Ms. Chiara Giglio, the Acquisitions Editor, and Ms. Tessa Kathryn, the Editorial Project Manager, and Mr. Sujithkumar Chandran, the Project Manager, for their unwavering support and meticulous efforts, which have been instrumental in ensuring the successful publication of this book.

CHAPTER 1

Introduction

The manufacturing industry is currently undergoing a significant transformation spurred by several key trends, including digitalization, sustainability, and personalization. These trends are reshaping the entire product lifecycle, spanning from design to production and final customer delivery. In this context, Human-Centric Smart Manufacturing (HCSM) represents an advanced manufacturing paradigm that capitalizes on cutting-edge technologies to optimize production processes, minimize waste, and enhance overall efficiency. To achieve it, Human–Robot Collaboration (HRC) stands out as a pivotal solution of HCSM, embodying a design philosophy centered around human involvement to establish a collaborative and adaptable production environment. It combines the unique strengths of both humans, characterized by cognitive flexibility and adaptability, and robots, distinguished by exceptional precision, strength, and repeatability. To gain a deeper insight into the evolution of HRC, a foundational understanding of the prevailing shift toward HCSM environments is essential, serving as a prerequisite for delving further into this subject.

1.1 Transition toward human-centric smart manufacturing

Today's manufacturing paradigm is evolving rapidly, driven by changing demands (Zheng et al., 2021). The proliferation of machining processes, product variations, and increased complexity has led to a growing need for human manual dexterity in various manufacturing tasks. Traditional large-scale production lines are becoming less effective and efficient in handling these evolving demands (Wang, 2022). Concurrently, there is a rising consumer preference for customized, personalized, and sustainable products. Meeting these preferences necessitates a manufacturing environment that is more flexible and adaptable (Wang et al., 2021; Mourtzis et al., 2019).

This shift has propelled the concept of HCSM to the forefront of discussions in leading manufacturing countries. Initiatives such as the Industrial Internet Consortium in the USA (Sisinni et al., 2018; Wang et al., 2020), Industry 5.0 in the European Commission (Breque et al., 2021), and Intelligent Manufacturing toward the year 2035 in China (Zhou et al., 2019) are gaining consensus. HCSM revolves around placing human workers at

Proactive Human–Robot Collaboration Toward Human-Centric
Smart Manufacturing
https://doi.org/10.1016/B978-0-44-313943-7.00008-9

1

the heart of the production process. It empowers humans to acquire new skills and tackle more complex tasks through the integration of advanced technologies like Artificial Intelligence (AI), Artificial General Intelligence (AGI), Augmented Reality (AR), and robotics. Simultaneously, HCSM prioritizes the well-being and job satisfaction of human workers by creating safer, more comfortable working conditions (Prati et al., 2021; Wang, 2019).

HRC plays a pivotal role in HCSM by fostering cooperation and coordination between human workers and robots within the production process (Leng et al., 2022). Human workers contribute unique skills, including decision-making, problem-solving, and creativity, whereas robots excel in speed, accuracy, and precision (Li et al., 2023). By capitalizing on the complementary strengths of both humans and robots, manufacturers can swiftly adapt to shifting customer demands, ensuring their competitiveness.

The market has witnessed the emergence of collaborative robots since 2015, such as Universal Robot's URx series, KUKA's LBR iiwa, ABB's YuMi, among others. These robots are designed to facilitate HRC, offering substantial potential for enhancing efficiency, productivity, and flexibility across a wide range of real-world manufacturing tasks.

For instance, in machining processes, HRC can provide a human operator with a Virtual Reality (VR) interface to better understand surface features and offer valuable decision feedback during precision grinding of mechanical components (Xie et al., 2022). In complex product assembly, HRC enables interactive task completion by fostering close, coordinated, and logical collaboration between humans and robots (Liu et al., 2022, 2019, 2021).

Efficient HRC hinges on enabling robots and humans to jointly perceive their shared workspace, make dynamic decisions, and collaborate while utilizing shared resources. Over the past decade, extensive research has delved into HRC across various production activities. By continuously monitoring changes within an overlapping workspace, collaborative robots can adapt their movements in real-time to ensure operator safety, especially in intricate assembly operations where close human–robot proximity is common (Polverini et al., 2017; Hietanen et al., 2020; Nikolakis et al., 2019). Furthermore, research has explored human action recognition to empower robots to make responsible decisions and swiftly execute supporting actions (Hammam et al., 2020). These investigations into human safety (Xu et al., 2021), scene perception (Fan et al., 2022), and robot adaptive control (Garcia et al., 2019) contribute to the wider acceptance of HRC in industrial applications (Li et al., 2021).

However, despite their potential advantages, current HRC systems exhibit critical weaknesses. Firstly, they often operate in a master/slave mode, limiting the full utilization of the complementary skills of both humans and robots. Secondly, conventional HRC systems primarily focus on perception capabilities, lacking knowledge-learning intelligence. Thirdly, these systems lack the capacity to anticipate future events, potentially leading to uncertain situations. Lastly, they heavily rely on predefined procedures, failing to exhibit active adaptability to evolving task requirements. In this context, the collaborative manufacturing system's efficiency is compromised as the full initiative of both humans and robots remains underutilized.

As a response to these challenges, the evolution of HRC is trending toward a proactive mode. Proactive HRC eliminates the need for humans and robots to compromise their performance to accommodate each other's leadership. Instead, it incorporates humanoid intelligence, the ability to anticipate spatio-temporal events, and self-coordination capabilities when confronted with complex environmental changes. In Proactive HRC, both humans and robots actively explore diverse solutions and manipulation techniques to successfully complete various manufacturing tasks, effectively addressing the limitations of current HRC systems.

This book, titled "Proactive Human–Robot Collaboration", represents a forward-looking paradigm within the context of human-centric smart manufacturing. Its primary objective is to systematically tackle the fundamental challenges associated with this concept through a technology-driven approach, thereby unlocking its considerable potential for adoption across industrial enterprises.

1.2 Motivation and vision

Before the effective implementation of Proactive HRC in real-world scenarios, the authors' primary motivation for writing this book is to address the following four fundamental questions:

Q1: How can future Proactive HRC systems transcend the limitations of the master/slave mode, fostering bi-directional and proactive collaboration while incorporating intelligent capabilities? What intellectual aptitudes can be cultivated in Proactive HRC to navigate environmental constraints, ensuring optimal human–robot performance within the dynamic HCSM setting?

Q2: Is it achievable for Proactive HRC to extract semantic knowledge regarding human–robot–task relationships and cultivate cognitive intelligence? How can empathetic teamwork skills be nurtured to align with the manipulation needs of both humans and robots, bridging the gap between ergonomic guidelines and robot control? Can Proactive HRC move beyond safety monitoring to actively respond to on-site scenario changes?

Q3: Can Proactive HRC proactively anticipate forthcoming events within the execution loop? How can mobile collaborative robots decipher "what the worker will do next" from partial video observations? Is it possible to establish a connection between robot planning and human intentions, enabling timely proactive responses rather than reactive ones?

Q4: Is it possibility for Proactive HRC to attain a comprehensive understanding of task fulfillment processes and autonomously allocate tasks based on evolving requirements, eschewing rigid, preprogrammed instructions? How can we effectively coordinate multiple human and robotic agents with diverse skills across shifting task stages, considering their distinct capabilities? Is it feasible to generate complex task arrangements within the framework of Proactive HRC in a transparent and comprehensible manner?

Driven by the previously mentioned challenges, this book sets out to delve into cutting-edge methodologies, including cognitive computing, AI, and robot learning, to propel the shift from conventional to Proactive HRC. The goal is to elevate its performance and expand its utility. This journey entails clarifying the essence of the Proactive HRC paradigm, elevating the system's cognitive capabilities, refining predictive planning, and facilitating self-organizing teamwork. These attributes will not only enhance the deployment of HRC systems but also provide insights into distinctive operational skills and common application scenarios.

On the other hand, the authors' motivation stems from their extensive dedication to the field of Proactive HRC, encompassing years of exploratory research endeavors. They enthusiastically embrace the opportunity to refine their accumulated knowledge and insights, transforming them into a comprehensive, step-by-step guide for Proactive HRC development, relevant to both academic and industrial contexts.

The book predominantly draws from a wealth of research projects conducted in Hong Kong, presenting a diverse array of novel findings and practical examples that vividly illustrate the distinctive aspects of Proactive

HRC. These insights are specifically tailored to facilitate the integration of Proactive HRC into industrial settings, with an emphasis on human-centric considerations. Furthermore, the authors suggest that the book can serve as a valuable reference for university students enrolled in advanced courses, particularly within the fields of industrial or mechanical engineering.

Ultimately, the authors aspire to ignite increased interest among researchers and practitioners in the Proactive HRC domain. They eagerly anticipate engaging in open discussions and fostering in-depth research and development efforts to further enhance the practical implementations of Proactive HRC in the era of human-centric smart manufacturing.

1.3 Content organization

This book, the inaugural publication in this field, offers a comprehensive exploration of Proactive HRC. It encompasses core concepts, definitions, distinguishing features, challenges, technical methodologies, deployment strategies, operational modes, industrial case studies, and potential avenues for future research and development. The book's structure is presented in Fig. 1.1, comprising a total of 10 chapters. Chapter 1 serves as the introductory overview, while Chapter 10 provides concluding remarks and highlights for the future. The remaining chapters are devoted to addressing the four pivotal questions (Q1–Q4) and providing detailed illustrations of Proactive HRC systems.

Chapter 2 delves into the typical relations between humans and robots in manufacturing, including coexistence, interaction, cooperation, and collaboration within today's HCSM environment. This distinction sets Proactive HRC apart from other HRC paradigms.

Chapter 3 further elucidates the essence of Proactive HRC, outlining its systematic architecture, cognitive capabilities, and enabling technologies that render it unique.

Chapters 4–6 explore state-of-the-art technical methodologies to imbue Proactive HRC with mutual-cognitive abilities, predictive capabilities, and self-organizing prowess. Chapter 4 introduces advanced visual reasoning techniques and Mixed Reality (MR) technologies, enabling empathic collaboration. Chapter 5 investigates multimodal deep learning algorithms and achieves ongoing human intention prediction and proactive robotic decision-making for predictable collaboration in the near future. Chapter 6 proposes a temporal subgraph reasoning-based approach for self-organizing HRC task planning among multiple agents.

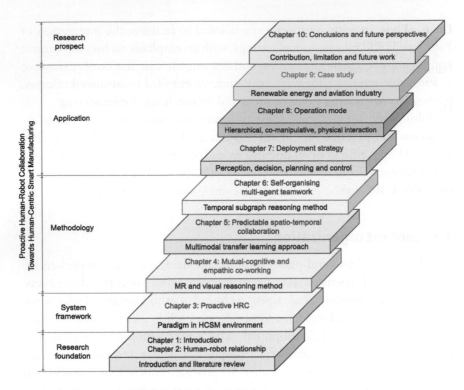

Figure 1.1 Content organization of this book.

Chapter 7 delineates step-by-step processes, from perception to decision, planning, and control levels, essential for the efficient deployment of a Proactive HRC system.

Chapter 8 elaborates on the typical operational modes of Proactive HRC, highlighting its unique skills in hierarchical operations, comanipulation, and physical interaction.

Chapter 9 offers two case studies showcasing Proactive HRC's application in manufacturing tasks, encompassing the renewable energy sector and the high–tech aviation industry.

References

Breque, M., De Nul, L., Petridis, A., 2021. Industry 5.0: Towards a Sustainable, Human-Centric and Resilient European Industry. European Commission, Directorate-General for Research and Innovation, Luxembourg.

Fan, J., Zheng, P., Li, S., 2022. Vision-based holistic scene understanding towards proactive human–robot collaboration. Robotics and Computer-Integrated Manufacturing 75, 102304.

Garcia, M.A.R., Rojas, R., Gualtieri, L., Rauch, E., Matt, D., 2019. A human-in-the-loop cyber-physical system for collaborative assembly in smart manufacturing. Procedia CIRP 81, 600–605.

Hammam, A.A., Soliman, M.M., Hassanien, A.E., 2020. Real-time multiple spatiotemporal action localization and prediction approach using deep learning. Neural Networks 128, 331–344.

Hietanen, A., Pieters, R., Lanz, M., Latokartano, J., Kämäräinen, J.K., 2020. AR-based interaction for human–robot collaborative manufacturing. Robotics and Computer-Integrated Manufacturing 63, 101891.

Leng, J., Sha, W., Wang, B., Zheng, P., Zhuang, C., Liu, Q., Wuest, T., Mourtzis, D., Wang, L., 2022. Industry 5.0: Prospect and retrospect. Journal of Manufacturing Systems 65, 279–295.

Li, S., Fan, J., Zheng, P., Wang, L., 2021. Transfer learning-enabled action recognition for human–robot collaborative assembly. Procedia CIRP 104, 1795–1800.

Li, S., Zheng, P., Liu, S., Wang, Z., Wang, X.V., Wang, L., 2023. Proactive human–robot collaboration: Mutual-cognitive, predictable, and self-organising perspectives. Robotics and Computer-Integrated Manufacturing 81, 102510.

Liu, Q., Liu, Z., Xu, W., Tang, Q., Zhou, Z., Pham, D.T., 2019. Human–robot collaboration in disassembly for sustainable manufacturing. International Journal of Production Research 57, 4027–4044.

Liu, S., Wang, L., Wang, X.V., 2021. Function block-based multimodal control for symbiotic human–robot collaborative assembly. Journal of Manufacturing Science and Engineering 143, 091001.

Liu, X., Zheng, L., Wang, Y., Yang, W., Jiang, Z., Wang, B., Tao, F., Li, Y., 2022. Human-centric collaborative assembly system for large-scale space deployable mechanism driven by digital twins and wearable AR devices. Journal of Manufacturing Systems 65, 720–742.

Mourtzis, D., Zogopoulos, V., Xanthi, F., 2019. Augmented reality application to support the assembly of highly customized products and to adapt to production re-scheduling. The International Journal of Advanced Manufacturing Technology 105, 3899–3910.

Nikolakis, N., Maratos, V., Makris, S., 2019. A cyber physical system (CPS) approach for safe human–robot collaboration in a shared workplace. Robotics and Computer-Integrated Manufacturing 56, 233–243.

Polverini, M.P., Zanchettin, A.M., Rocco, P., 2017. A computationally efficient safety assessment for collaborative robotics applications. Robotics and Computer-Integrated Manufacturing 46, 25–37.

Prati, E., Peruzzini, M., Pellicciari, M., Raffaeli, R., 2021. How to include user experience in the design of human–robot interaction. Robotics and Computer-Integrated Manufacturing 68, 102072.

Sisinni, E., Saifullah, A., Han, S., Jennehag, U., Gidlund, M., 2018. Industrial Internet of Things: Challenges, opportunities, and directions. IEEE Transactions on Industrial Informatics 14, 4724–4734.

Wang, B., Li, X., Freiheit, T., Epureanu, B.I., 2020. Learning and intelligence in human–cyber-physical systems: Framework and perspective. In: 2020 Second International Conference on Transdisciplinary AI (TransAI), IEEE, pp. 142–145.

Wang, L., 2019. From intelligence science to intelligent manufacturing. Engineering 5, 615–618.

Wang, L., 2022. A futuristic perspective on human-centric assembly. Journal of Manufacturing Systems 62, 199–201.

Wang, Z., Chen, C.H., Zheng, P., Li, X., Khoo, L.P., 2021. A graph-based context-aware requirement elicitation approach in smart product-service systems. International Journal of Production Research 59, 635–651.

Xie, H.L., Wang, Q.H., Ong, S., Li, J.R., Chi, Z.P., 2022. Adaptive human–robot collaboration for robotic grinding of complex workpieces. CIRP Annals.

Xu, W., Cui, J., Li, L., Yao, B., Tian, S., Zhou, Z., 2021. Digital twin-based industrial cloud robotics: Framework, control approach and implementation. Journal of Manufacturing Systems 58, 196–209.

Zheng, P., Xia, L., Li, C., Li, X., Liu, B., 2021. Towards self-x cognitive manufacturing network: An industrial knowledge graph-based multi-agent reinforcement learning approach. Journal of Manufacturing Systems 61, 16–26.

Zhou, Y., Zang, J., Miao, Z., Minshall, T., 2019. Upgrading pathways of intelligent manufacturing in China: Transitioning across technological paradigms. Engineering 5, 691–701.

CHAPTER 2

Evolution of human–robot relationships

Since the integration of industrial robots into extensive production lines, there has been a significant emphasis among researchers and engineers on the development of human–robot relationships. As manufacturing needs to evolve, human–robot relations progress through distinct stages to enhance productivity. In this chapter, we review the evolution of human–robot relationships, transitioning from Coexistence and Interaction to Cooperation and Collaboration. We also introduce the 5C intelligence framework, encompassing Connection, Coordination, Cyber, Cognition, and Coevolution. Within this framework, we highlight the emergence of Proactive HRC as a prospective paradigm that aligns with the human-centric principles of Industry 5.0.

2.1 Human–robot coexistence

In the distant past, dating back to 1979, a select group of researchers delved into the research of modeling robots as operators (Paul and Nof, 1979). This exploration marked the inception of human–robot coexistence within the field of manufacturing tasks. At its core, this coexistence entails the spatial coplacement of both a robot and a human, carefully arranged in such a way that their respective workspaces remain nonoverlapping. A notable characteristic of this scenario is the absence of direct physical contact between the human and the robot. The potential exists for the exchange of the work object between these entities, although this exchange transpires independently and concurrently. The demonstration of human–robot coexistence in manufacturing is shown in Fig. 2.1.

As shown in Table 2.1, in this stage, Kamali et al. (1982) introduced a framework for selecting suitable robots, machines, and conveyors to complement human workers, demonstrated effectively through a transaxle assembly task, showcasing the approach's efficiency in optimizing systems. Argote et al. (1983) investigated worker responses to robot introductions in factories, with a focus on their psychological reactions. Workers reported benefits such as reduced fatigue, but also drawbacks like increased down-

Proactive Human–Robot Collaboration Toward Human-Centric
Smart Manufacturing
https://doi.org/10.1016/B978-0-44-313943-7.00009-0

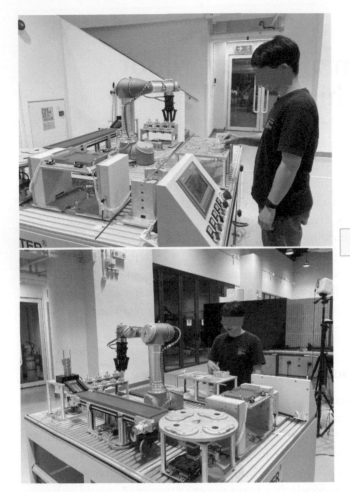

Shared Simultaneous process

Figure 2.1 Demonstration of human–robot coexistence in manufacturing.

time. Additionally, Wakita et al. (2001) explored the importance of efficient information sharing between robots and humans for successful coexistence, exemplifying its effectiveness through an implemented projection system. In pursuit of coexistence, Lam et al. (2010) developed an algorithm called "human–centered sensitive navigation", which takes into account the states of both humans and robots. This approach introduces coexistence rules prioritizing safety and smooth navigation, while recognizing the presence of sensitive zones around both humans and robots to encourage socially acceptable movements. Furthermore, addressing safety concerns in human–

Table 2.1 Examples of human–robot coexistence in industry settings.

Objective	Method	Reference
Integration humans and robots in production	Task alternation considering human and robot characteristics	Kamali et al. (1982)
Impact of introduction of robots besides humans	Milling processes and probit analysis	Argote et al. (1983)
Information sharing between robots and humans	Video projector and CCD camera	Wakita et al. (2001)
Safety and smooth navigation of robots	Finite-state machine and collision-free motion planner	Lam et al. (2010)
Safety problems in coexistence	Distance safety index and momentum safety index	Tsai et al. (2014)

robot coexistence, Tsai et al. (2014) introduced an optimization-based path planning framework that uses ellipsoid coordinates and accounts for inertial effects. It generates real-time, collision-free trajectories for robots, thus reducing computational load and validation through simulations involving two different robots.

2.2 Human–robot interaction

Leveraging sensor and communication technology, human–robot relationships progressed to incorporate intelligent capabilities, marking the initial stage of interaction as highlighted in Ghosh and Helander (1986). Human–robot interaction (see Fig. 2.2) occurs when a human and a robot coexist within the same physical workspace and engage in mutual communication. This interaction can take various forms, such as one party providing guidance or exerting control over the other, or when there is any form of physical contact, whether intentional or unintentional, as presented in Table 2.2. Notably, both the human and the robot collaborate on a shared task, progressing through each step in a sequential manner.

In this phase, Kulić and Croft (2005) introduced a strategy aimed at enhancing the safety of human–robot interactions by focusing on hand-off tasks between articulated robots and inexperienced human users. The study presented two distinct formulations of the danger criterion, one taking into account independent safety factors and the other considering mutually dependent factors. Fritzsche et al. (2011), in the context of safe physical

Figure 2.2 Demonstration of human–robot interaction in manufacturing.

interaction within shared workspaces, developed a pressure-sensitive artificial skin with integrated cushioning elements. This innovation not only
enabled reliable contact measurements across the entire robot body but also
facilitated safety-related functions and touch-based robot motion control,
simplifying human–robot interactions and mitigating the risk of injuries.
For bilateral interaction, Leica et al. (2013) concentrated on recognizing
human intent through mechanical impedance modeling, which guided the
robot through unstructured environments while considering robot dynamics, as analyzed using Lyapunov theory.

Table 2.2 Examples of human–robot interaction in industry settings.

Objective	Method	Reference
Safety problems in interaction	A sum–based and a product–based criterion for control	Kulić and Croft (2005)
Safe physical interaction	Robot pressure-sensitive skin	Fritzsche et al. (2011)
Bilateral interaction with physical contact	Lyapunov theory-based robot kinematic controller	Leica et al. (2013)
Intuitive and efficient interface	Euclidean distance-based method for workspace location in AR interface	Fang et al. (2014)
Communication between humans and robots	Gaze pattern detection and robot interactive actions	Das et al. (2014)
Communication between multiple humans and robots	Multimodal information fusion, hand sign recognition and emotion recognition	Luo et al. (2015)
Trust issue in interaction	Independent samples' t-test and Levene's test for equality of variances	Lazányi and Hajdu (2017)
Nonverbal communication	posture and gesture recognition	Stoeva and Gelautz (2020)

In pursuit of intuitive and efficient interfaces for human–robot interaction, Fang et al. (2014) introduced an AR-based interface. This approach offered diverse interaction methods suitable for various robotic applications, employing a Euclidean distance-based approach and monitor-based visualization. It streamlined tasks like robot pick-and-place operations and path following. Furthermore, Das et al. (2014) presented a human–robot interaction system capable of detecting and classifying a target human's gaze pattern. Based on this pattern, the robot could seamlessly respond to humans. Luo et al. (2015) adopted a multimodal approach, fusing information on hand signs and emotions for communication needs in human–robot interactions. The integrated system had the ability to track multiple individuals, identify their facial expressions and emotional states, enabling robots to respond effectively to different people. In the context of trust in interaction, Lazányi and Hajdu (2017) proposed the role of trust, considering both dispositional and historical records. The study highlighted the importance of dispositional trust among young adults while identifying factors influencing historical trust. Lastly, Stoeva and Gelautz (2020) delved into nonverbal communication methods for human–robot interaction, particularly focusing on body movements, postures, and gestures as a means of conveying affective information.

2.3 Human–robot cooperation

Through the integration of scenario perception and optimal controllers, human–robot agents gained a degree of autonomy within the execution loop, contributing to the advancement of human–robot cooperation (Laengle et al., 1997). This cooperative environment allows human and robot agents, each pursuing their distinct goals and objectives, to temporarily share physical, cognitive, or computational resources for mutual benefit. While the parties may operate within a partially overlapping workspace, it is noted that direct physical contact is not the norm, as they can work concurrently or may occasionally have to wait for the availability of other agents, as presented in Table 2.3. The typical scenario of human–robot cooperation in industrial settings is shown in Fig. 2.3.

In this stage, Gecks and Henrich (2005) introduced a system that involved an industrial robot working alongside a human to carry a box. This setup utilized an image-based method for obstacle detection, allowing the robot to adjust its motion path in real-time when collisions were identified, enabling safe and efficient pick-and-place operations. To cope with various environmental and task constraints in human–robot cooperation, Seto et al. (2007) developed a real-time self-collision avoidance motion generation method for a mobile manipulator. This method considered the robot's joint movement range and was applied to a human-friendly robot, demonstrating its effectiveness through experimental validation.

In the context of construction tasks involving heavy materials, Cherubini et al. (2013) designed a human–robot cooperative system that combined a multi-DoF manipulator with a miniexcavator on construction sites. This system facilitated real-time interactions among humans, robots, and the environment, enhancing the handling of heavy construction materials while enabling humans to perceive and respond to forces exerted by the robot's end effector.

Subsequently, Lee and Song (2014) proposed a multimodal sensor-based control framework to enhance human–robot cooperation. This approach used a Kinect and an onboard camera based on a unified task formalism, and it was successfully validated in a simulated industrial setting where

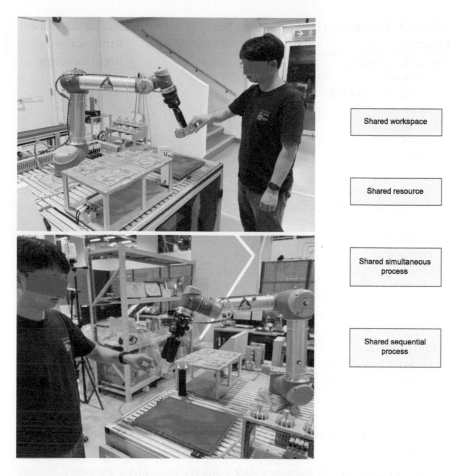

Figure 2.3 Demonstration of human–robot cooperation in manufacturing.

humans and robots collaborated in inserting screws into a flank. In the context of collision risk mitigation during human–robot cooperation, Whitsell and Artemiadis (2015) introduced a collision detection method capable of distinguishing unintended collisions from intended task forces at the robot's end effector. This approach was experimentally verified, proving its effectiveness even when physical contact occurred between humans and robots.

To address role switching in human–robot cooperation, Whitsell and Artemiadis (2015) explored a novel approach that allowed for the simultaneous switching of two roles during translation tasks. Regardless of whether fixed or adaptive parameters were used for role switching, the method facili-

Table 2.3 Examples of human–robot cooperation in industry settings.

Objective	Method	Reference
Safe pick-and-place operation	Image update algorithm	Kulić and Croft (2005)
Environmental and task constraint	Self-collision avoidance motion generation	Seto et al. (2007)
Heavy material installing	Impedance control strategy in cooperation	Lee et al. (2007)
Intuitive cooperation	State machine for selecting the appropriate control mode	Cherubini et al. (2013)
Collision avoidance during cooperation	A linear mapping between the joint torque space and the force space	Lee and Song (2014)
Role switching during cooperation	Leader/follower control strategy	Whitsell and Artemiadis (2015)
Automatic generation of a robot program during cooperation	CAD-based task and world model, task distribution planner, robot program generator, and an action recognition module	Berg and Reinhart (2017)
Adaptive robotic behavior considering human motion limits	Linear quadratic regulation and integral reinforcement learning	Li et al. (2017)

tated successful human–robot cooperation. Berg and Reinhart (2017) tackled the challenges of task distribution complexities and time-consuming robot programming in human–robot cooperation. Their proposed system combined automated task distribution between humans and robots with a task-oriented programming system, simplifying task allocation through a user interface and automatically generating robot programs with assembly task modules. These modules relied on camera-based detection of hand and object movements using Hidden Markov Models.

Lastly, Li et al. (2017) introduced a human–robot cooperation system with adaptive robotic behavior, designed to enhance task performance while considering human arm motion limits. Their approach included an adaptive impedance control method for robotic manipulators, addressing motion tracking errors and optimizing impedance modes through Linear Quadratic Regulation (LQR) and Integral Reinforcement Learning (IRL). This system effectively assisted operators in human–robot cooperative tasks.

2.4 Human–robot collaboration

Since 2008, HRC (Bi et al., 2008) has taken on an increasingly pivotal role, facilitating highly flexible manufacturing in line with the growing emphasis on personalized production. Collaboration, in this context, represents the co-working in manufacturing tasks involving both humans and robots operating within a shared workspace, collectively achieving a specific set of assigned objectives. This collaborative process necessitates a synchronized and coordinated effort from all participants, wherein physical contact between humans and robots is also permissible. In HRC, human and robots share their distinct capabilities, competences, and resources towards the shared goal in joint manufacturing activities. Fig. 2.4 presents the human–robot collaborative assembly in close proximity.

In Table 2.4, De Tommaso et al. (2012) improved robot skill acquisition in HRC by introducing an active interface, enabling seamless interaction in the shared workspace with real-time visual feedback. Meanwhile, Charalambous et al. (2015) delved into the analysis of human factors influencing the successful implementation of industrial HRC, emphasizing the need to pay attention to these factors.

In collaborative assembly, Wang et al. (2017) underscored the industry's shift toward safe and reliable robotic collaboration with humans. They introduced a comprehensive framework for a human–robot collaborative assembly system, substantiated by a case study in cyber-physical production. For HRC involving a human and a dual-arm robot in manufacturing, Coupeté et al. (2019) developed a real-time gesture recognition algorithm, facilitating a smooth, natural, and efficient coordination between robot and human operators. Furthermore, Wang et al. (2018) tackled the challenge of human–robot object handover in manufacturing. They proposed a practical approach that employs a wearable sensory system to enhance the efficiency and naturalness of the handover process in manufacturing environments by allowing robots to recognize human intentions and attributes of objects.

In pursuit of intuitive HRC, Wang et al. (2020) developed a feasible AR-based system, implemented on the HoloLens device with a closed-loop structure. This system improved response times and compensation accuracy for effective collaboration. Addressing the issue of safety assessment in HRC, Liu et al. (2020) established an assessment model and introduced a dynamic modified Speed and Separation Monitoring (SSM) method, ensuring safety while optimizing productivity. They validated this approach through a prototype system with real-time risk assessment and safe motion

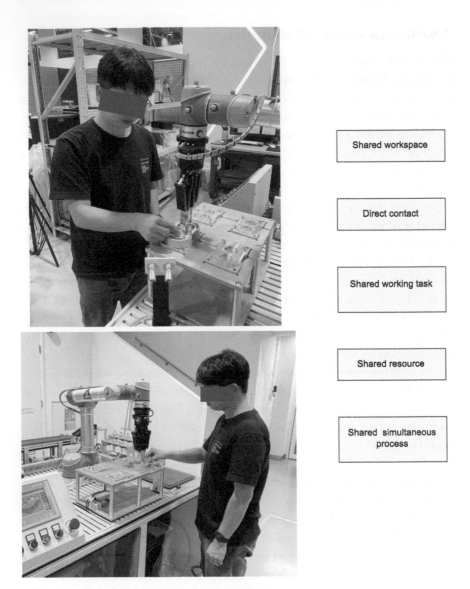

Shared workspace

Direct contact

Shared working task

Shared resource

Shared simultaneous process

Figure 2.4 Demonstration of human–robot collaboration in manufacturing.

control, demonstrating its effectiveness in an HRC cell with an industrial robot.

In construction sites, where robots often require human assistance, Liu et al. (2021b) introduced a worker-centered collaborative framework that used wearable electroencephalograph technology to capture workers' brain-

Table 2.4 Examples of HRC in industry settings.

Objective	Method	Reference
Interactive skill teaching and collaborative assembly	Learning from demonstration and mixed reality-based interface	De Tommaso et al. (2012)
Human factor during collaboration	Template Analysis	Charalambous et al. (2015)
Human–robot collaborative assembly	Cyber-physical systems and cloud based	Wang et al. (2017)
Collaboration between a human and a dual-arm robot	Online gesture recognition and adaptation to new users	Coupeté et al. (2019)
Human–robot object handover	Human hand intentions detected by EMG signals	Wang et al. (2018)
AR-based HRC with accurate robot control	AR layer, virtual environment, robot command and robot controller	Wang et al. (2020)
Dynamic risk assessment in HRC	ISO/TS 15066 oriented dynamic risk assessment model	Liu et al. (2020)
Brainwave-driven HRC	Near-real-time signal decoding, Fault-tolerant mechanism for robotic command generation	Liu et al. (2021b)
Dynamic task scheduling for HRC	Induced subgraph	Alirezazadeh and Alexandre (2022)
Flexible scheduling and tactile communication in HRC	AND/OR graph, haptic ring, and haptic bracelets	Maderna et al. (2022)

waves. The robot adapted its performance based on their cognitive load, accurately adjusting its working pace and fostering trust and safety in HRC. For dynamic task allocation and scheduling in HRC systems, Alirezazadeh and Alexandre (2022) concentrated on task precedence order and real-time monitoring of human agent performance. Their method automatically prioritized tasks, assigned them to the appropriate agent based on quality metrics, and managed task transfers as human performance changed. Finally, in response to scheduling algorithms in flexible manufacturing HRC, Maderna et al. (2022) introduced a dynamic scheduler that adapted to system variability, which can communicate instructions to human operators through haptic guidance.

2.5 From HRC to Proactive HRC

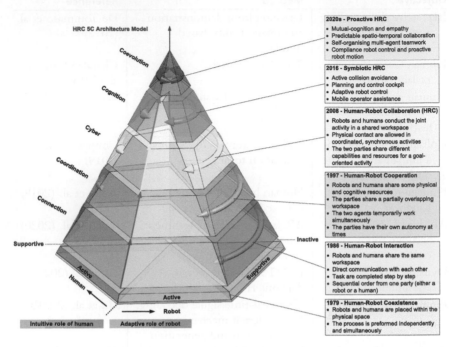

Figure 2.5 An evolvement pathway of human–robot relationships toward Proactive HRC. (Adapted from Li et al. (2023).)

Fig. 2.5 illustrates the evolution of human–robot relationships toward Proactive HRC through six distinct phases, characterized by the degree of complementarity between the two parties (horizontal axis) and their level of intelligent capabilities (vertical axis). The horizontal axis, depicted as a hexagon, signifies the collaborative engagement and shared responsibility between humans and robots throughout a given task (Wang et al., 2019). Within the various phases of task execution, human and robotic agents may assume different roles, including an active role as the primary decision-maker, a supportive role performing tasks with assistance, and an inactive role, where they observe and trust their counterpart's actions. This adaptability in roles, guided by intuitive human behaviors and adaptive robot control, allows both parties to dynamically adjust their roles to accommodate changing task requirements:

In parallel, the vertical axis represents the levels of smartness in human–robot relationships, as derived from the 5C architecture of the Cyber-

Physical System (CPS) model (Lee et al., 2015). These levels encompass various stages of collaborative capabilities, including:

i) (Connection) Reflecting the ability for humans and robots to jointly perform work in parallel, such as coassembling a gearbox under predefined procedures.

ii) (Coordination) Signifying the capacity for simultaneous cooperation based on sensorial and perceptual feedback from the surrounding environment, such as robots detecting obstacles using visual sensors.

iii) (Cyber) Denoting synchronous and dynamic collaboration among all parties within a previewable and predictable execution loop, potentially involving cloud robotics and shared resource utilization.

iv) (Cognition) Representing the cognitive understanding and high-level decision-making capabilities within human–robot organizations, including human intention recognition and robot learning.

v) (Coevolution) Expressing self-fulfillment goals and collaborative intelligence, such as self-organizing resource allocation and ergonomic cooperation.

Each higher level of smartness inherits and accumulates characteristics from the underlying architectures. The unattained degree of active behaviors for human and robotic agents is indicated within the shaded gray block within each smartness level. To provide a comprehensive understanding of each paradigm, their critical characteristics are summarized in the respective box, and their evaluation results are highlighted in a three-dimensional coordinate system accordingly.

As delineated in the preceding sections, in the initial phase of human–robot coexistence, both humans and robots operate within the same workspace but in a rather disjointed manner, essentially carrying out their designated tasks while remaining segregated by physical barriers, as illustrated by the black arrows and the delineated black dashed line in Fig. 2.5. At this stage, there is an absence of true collaborative awareness between humans and robots.

Subsequently, as the domain of human–robot interaction matures, it permits seamless communication and joint work between these two agents (Hara and Kobayashi, 1996). Within this phase, humans and robots demonstrate an inclination to follow orders from their counterparts, completing operation sequences in a sequential manner (Friedrich et al., 1996). Although more sophisticated interaction modalities emerge over time, including the adoption of physical haptic interfaces (Abbink et al., 2012), gestures as a means of communication (Nickel and Stiefelhagen, 2007), and even brain–computer interfaces (Ferrez and Millán, 2008), the prevailing theme is one of a partner providing supportive behaviors to assist the other, rather than achieving true mutual active synergism, as indicated by the blue arrow in Fig. 2.5.

The evolving relationship within the realm of human–robot cooperation is symbolized by the pink arrow in Fig. 2.5. During this phase, research endeavors concentrated on applications in industrial robotics, particularly in domains like pick-and-place operations (Gecks and Henrich, 2005), heavy material installation (Lee et al., 2007), and efficient object handover mechanisms (Ikeura et al., 2002). These collective efforts have significantly propelled the adoption of flexible manufacturing practices since the onset of the mid-2000s (Stopp et al., 2005; Koeppe et al., 2005).

In the HRC stage, technologies from CPS (Lee, 2008) and robot decision-making (Cao et al., 2008) empower humans and robots within HRC systems to share their distinct capabilities and on-demand resources throughout the execution loop. However, the challenge of this stage lies in achieving a comprehensive, high-level understanding of manufacturing tasks, thus limiting the promotion of active-role behaviors in this collaborative environment, as indicated by the yellow arrow in Fig. 2.5. Notably, much of the research within this stage has concentrated on nonsemantic perceptual aspects, such as human motion estimation (Li and Ge, 2013), operator stress assessment (Arai et al., 2010), and the development of safety control strategies (Tan et al., 2010).

Transitioning to the next phase, the Symbiotic HRC paradigm, marked by the green arrow, leverages cognitive computing and multimodal communication techniques (Liu et al., 2021a) to enhance manufacturing performance by synergizing the complementary competencies of human and robotic agents (Wang et al., 2019). This approach includes four key elements: active collision avoidance, a planning and control cockpit, adaptive robot control, and mobile operator assistance. However, despite these advancements, contemporary HRC development grapples with shortcomings that hinder the effective collaboration of robotic automation and human cognition. First, the collaboration between humans and robots often remains unidirectional and constrained within a slave/master framework, leaning either toward a human- or robot-leading approach, which restricts adaptability and flexibility. Furthermore, limited research has explored the concept of a Proactive HRC system, which possesses the intellectual capacity to acquire manufacturing knowledge and make flexible decisions akin to a human. Additionally, there has been a lack of discourse concerning the notion of mutual-cognitive, predictable, and self-organizing intelligence that could foster proactive behaviors in both humans and robots.

To tackle these challenges, in light of ongoing research in human-level information processing through cognitive computing, the Industrial Internet of Things, and robot learning, a human-centric smart manufacturing paradigm known as "Proactive HRC" has been introduced and defined by Li et al. (2023). Proactive HRC is described as "*a knowledge-driven, intelligent collaboration between human operators and robots, rooted in a comprehensive understanding of the on-site environment, transparent decision-making for resource allocation, and adaptable task execution in the face of industrial uncertainties.*" This paradigm represents the culmination of the Symbiotic HRC trajectory, incorporating a high level of automation, enabling bi-directional collaboration between humans and robots in manufacturing activities.

As we observe the progression of smartness illustrated by the purple arrow in Fig. 2.5, advanced Proactive HRC comprises four essential modules. Firstly, it fosters mutual cognition and empathy within the human–robot–workspace execution loop. Secondly, it promotes predictable spatio-temporal collaboration, enhancing the efficiency of task completion. Thirdly, it facilitates self-organizing multi-agent teamwork with dynamic resource allocation. Finally, it implements compliance robot control and proactive robot motion. Proactive HRC is designed to align with human-centric requirements, aiming to achieve the optimal blend of human and robot intelligent capabilities, thereby enhancing overall productivity and product quality.

References

Abbink, D.A., Mulder, M., Boer, E.R., 2012. Haptic shared control: Smoothly shifting control authority? Cognition, Technology & Work 14, 19–28.

Alirezazadeh, S., Alexandre, L.A., 2022. Dynamic task scheduling for human–robot collaboration. IEEE Robotics and Automation Letters 7, 8699–8704.

Arai, T., Kato, R., Fujita, M., 2010. Assessment of operator stress induced by robot collaboration in assembly. CIRP Annals 59, 5–8.

Argote, L., Goodman, P.S., Schkade, D., 1983. The human side of robotics: How worker's react to a robot. Sloan Management Review.

Berg, J., Reinhart, G., 2017. An integrated planning and programming system for human–robot cooperation. Procedia CIRP 63, 95–100.

Bi, Z., Lang, S.Y., Wang, L., 2008. Improved control and simulation models of a tricycle collaborative robot. Journal of Intelligent Manufacturing 19, 715–722.

Cao, M., Stewart, A., Leonard, N.E., 2008. Integrating human and robot decision-making dynamics with feedback: Models and convergence analysis. In: 2008 47th IEEE Conference on Decision and Control, IEEE, pp. 1127–1132.

Charalambous, G., Fletcher, S., Webb, P., 2015. Identifying the key organisational human factors for introducing human–robot collaboration in industry: An exploratory study. The International Journal of Advanced Manufacturing Technology 81, 2143–2155.

Cherubini, A., Passama, R., Meline, A., Crosnier, A., Fraisse, P., 2013. Multimodal control for human–robot cooperation. In: 2013 IEEE/RSJ International Conference on Intelligent Robots and Systems, IEEE, pp. 2202–2207.

Coupeté, E., Moutarde, F., Manitsaris, S., 2019. Multi-users online recognition of technical gestures for natural human–robot collaboration in manufacturing. Autonomous Robots 43, 1309–1325.

Das, D., Rashed, M.G., Kobayashi, Y., Kuno, Y., 2014. Recognizing gaze pattern for human–robot interaction. In: Proceedings of the 2014 ACM/IEEE International Conference on Human–Robot Interaction, pp. 142–143.

De Tommaso, D., Calinon, S., Caldwell, D.G., 2012. A tangible interface for transferring skills: Using perception and projection capabilities in human–robot collaboration tasks. International Journal of Social Robotics 4, 397–408.

Fang, H., Ong, S.K., Nee, A.Y., 2014. Novel AR-based interface for human–robot interaction and visualization. Advances in Manufacturing 2, 275–288.

Ferrez, P.W., Millán, J.d.R., 2008. Error-related EEG potentials generated during simulated brain–computer interaction. IEEE Transactions on Biomedical Engineering 55, 923–929.

Friedrich, H., Mnch, S., Dillmann, R., Bocionek, S., Sassin, M., 1996. Robot programming by demonstration (RPD): Supporting the induction by human interaction. Machine Learning 23, 163–189.

Fritzsche, M., Elkmann, N., Schulenburg, E., 2011. Tactile sensing: A key technology for safe physical human robot interaction. In: Proceedings of the 6th International Conference on Human–Robot Interaction, pp. 139–140.

Gecks, T., Henrich, D., 2005. Human–robot cooperation: safe pick-and-place operations. In: ROMAN 2005. IEEE International Workshop on Robot and Human Interactive Communication. IEEE, pp. 549–554.

Ghosh, B.K., Helander, M.G., 1986. A systems approach to task allocation of human–robot interaction in manufacturing. Journal of Manufacturing Systems 5, 41–49.

Hara, F., Kobayashi, H., 1996. Real-time facial interaction between human and 3D face robot agent. In: Proceedings 5th IEEE International Workshop on Robot and Human Communication. RO-MAN'96 TSUKUBA. IEEE, pp. 401–409.

Ikeura, R., Moriguchi, T., Mizutani, K., 2002. Optimal variable impedance control for a robot and its application to lifting an object with a human. In: Proceedings. 11th IEEE International Workshop on Robot and Human Interactive Communication. IEEE, pp. 500–505.

Kamali, J., Moodie, C.L., Salvendy, G., 1982. A framework for integrated assembly systems: Humans, automation and robots. International Journal of Production Research 20, 431–448.

Koeppe, R., Engelhardt, D., Hagenauer, A., Heiligensetzer, P., Kneifel, B., Knipfer, A., Stoddard, K., 2005. Robot–robot and human–robot cooperation in commercial robotics applications. In: Robotics Research. The Eleventh International Symposium. Springer, pp. 202–216.

Kulić, D., Croft, E.A., 2005. Safe planning for human–robot interaction. Journal of Robotic Systems 22, 383–396.

Laengle, T., Hoeniger, T., Zhu, L., 1997. Cooperation in human–robot teams. In: ISIE'97 Proceeding of the IEEE International Symposium on Industrial Electronics. IEEE, pp. 1297–1301.

Lam, C.P., Chou, C.T., Chiang, K.H., Fu, L.C., 2010. Human-centered robot navigation—towards a harmoniously human–robot coexisting environment. IEEE Transactions on Robotics 27, 99–112.

Lazányi, K., Hajdu, B., 2017. Trust in human–robot interactions. In: 2017 IEEE 14th International Scientific Conference on Informatics. IEEE, pp. 216–220.

Lee, E.A., 2008. Cyber physical systems: Design challenges. In: 2008 11th IEEE International Symposium on Object and Component-Oriented Real-Time Distributed Computing (ISORC). IEEE, pp. 363–369.

Lee, J., Bagheri, B., Kao, H.A., 2015. A cyber-physical systems architecture for Industry 4.0-based manufacturing systems. Manufacturing Letters 3, 18–23.

Lee, S.D., Song, J.B., 2014. Collision detection for safe human–robot cooperation of a redundant manipulator. In: 2014 14th International Conference on Control, Automation and Systems (ICCAS 2014). IEEE, pp. 591–593.

Lee, S.Y., Lee, K.Y., Lee, S.H., Kim, J.W., Han, C.S., 2007. Human–robot cooperation control for installing heavy construction materials. Autonomous Robots 22, 305–319.

Leica, P., Toibero, J.M., Roberti, F., Carelli, R., 2013. Bilateral human–robot interaction with physical contact. In: 2013 16th International Conference on Advanced Robotics (ICAR). IEEE, pp. 1–6.

Li, S., Zheng, P., Liu, S., Wang, Z., Wang, X.V., Wang, L., 2023. Proactive human–robot collaboration: Mutual-cognitive, predictable, and self-organising perspectives. Robotics and Computer-Integrated Manufacturing 81, 102510.

Li, Y., Ge, S.S., 2013. Human–robot collaboration based on motion intention estimation. IEEE/ASME Transactions on Mechatronics 19, 1007–1014.

Li, Z., Liu, J., Huang, Z., Peng, Y., Pu, H., Ding, L., 2017. Adaptive impedance control of human–robot cooperation using reinforcement learning. IEEE Transactions on Industrial Electronics 64, 8013–8022.

Liu, S., Wang, L., Wang, X.V., 2021a. Function block-based multimodal control for symbiotic human–robot collaborative assembly. Journal of Manufacturing Science and Engineering 143, 091001.

Liu, Y., Habibnezhad, M., Jebelli, H., 2021b. Brainwave-driven human–robot collaboration in construction. Automation in Construction 124, 103556.

Liu, Z., Wang, X., Cai, Y., Xu, W., Liu, Q., Zhou, Z., Pham, D.T., 2020. Dynamic risk assessment and active response strategy for industrial human–robot collaboration. Computers & Industrial Engineering 141, 106302.

Luo, R.C., Wu, Y., Lin, P., 2015. Multimodal information fusion for human–robot interaction. In: 2015 IEEE 10th Jubilee International Symposium on Applied Computational Intelligence and Informatics. IEEE, pp. 535–540.

Maderna, R., Pozzi, M., Zanchettin, A.M., Rocco, P., Prattichizzo, D., 2022. Flexible scheduling and tactile communication for human–robot collaboration. Robotics and Computer-Integrated Manufacturing 73, 102233.

Nickel, K., Stiefelhagen, R., 2007. Visual recognition of pointing gestures for human–robot interaction. Image and Vision Computing 25, 1875–1884.

Paul, R.P., Nof, S.Y., 1979. Work methods measurement—a comparison between robot and human task performance. International Journal of Production Research 17, 277–303.

Seto, F., Hirata, Y., Kosuge, K., 2007. Motion generation method for human–robot cooperation to deal with environmental/task constraints. In: 2007 IEEE International Conference on Robotics and Biomimetics (ROBIO). IEEE, pp. 646–651.

Stoeva, D., Gelautz, M., 2020. Body language in affective human–robot interaction. In: Companion of the 2020 ACM/IEEE International Conference on Human–Robot Interaction, pp. 606–608.

Stopp, A., Baldauf, T., Horstmann, S., Kristensen, S., 2005. Toward safe human–robot co-operation in manufacturing. In: Advances in Human–Robot Interaction. Springer, pp. 255–265.

Tan, J.T.C., Duan, F., Kato, R., Arai, T., 2010. Safety strategy for human–robot collaboration: Design and development in cellular manufacturing. Advanced Robotics 24, 839–860.

Tsai, C.S., Hu, J.S., Tomizuka, M., 2014. Ensuring safety in human–robot coexistence environment. In: 2014 IEEE/RSJ International Conference on Intelligent Robots and Systems. IEEE, pp. 4191–4196.

Wakita, Y., Hirai, S., Suehiro, T., Hori, T., Fujiwara, K., 2001. Information sharing via projection function for coexistence of robot and human. Autonomous Robots 10, 267–277.

Wang, L., Gao, R., Váncza, J., Krüger, J., Wang, X.V., Makris, S., Chryssolouris, G., 2019. Symbiotic human–robot collaborative assembly. CIRP Annals 68, 701–726.

Wang, W., Li, R., Diekel, Z.M., Chen, Y., Zhang, Z., Jia, Y., 2018. Controlling object hand-over in human–robot collaboration via natural wearable sensing. IEEE Transactions on Human–Machine Systems 49, 59–71.

Wang, X.V., Kemény, Z., Váncza, J., Wang, L., 2017. Human–robot collaborative assembly in cyber-physical production: Classification framework and implementation. CIRP Annals 66, 5–8.

Wang, X.V., Wang, L., Lei, M., Zhao, Y., 2020. Closed-loop augmented reality towards accurate human–robot collaboration. CIRP Annals 69, 425–428.

Whitsell, B., Artemiadis, P., 2015. On the role duality and switching in human–robot cooperation: An adaptive approach. In: 2015 IEEE International Conference on Robotics and Automation (ICRA). IEEE, pp. 3770–3775.

CHAPTER 3

Fundamentals of proactive human–robot collaboration

HRC plays a crucial role in smart manufacturing, addressing the demanding requirements of human-centricity, sustainability, and resilience. However, current HRC development tends to adopt either a human- or robot-dominant approach, wherein human and robotic agents reactively execute tasks based on predefined instructions. This falls short of achieving efficient integration between robotic automation and human cognition. These rigid human–robot relationships are ill-suited for intricate manufacturing tasks and do not alleviate the physical and psychological burdens on human operators. In response to these practical needs, this chapter presents our perspective on the foreseeable trend, concept, systematic framework, and enabling technologies of Proactive HRC. This forward-looking vision represents a compelling research avenue for future HCSM environment. The evolution of the human–robot symbiotic relationship is characterized by 5C intelligence, progressing from Connection, Coordination, Cyber, and Cognition to Coevolution. Ultimately, it culminates in mutual-cognitive, predictable, and self-organizing intelligent capabilities encapsulated in Proactive HRC. With proactive robot control, multiple human and robotic agents collaborate seamlessly in manufacturing tasks, taking into account each other's operational requirements, resource preferences, and complementary capabilities. This framework is envisioned to stimulate open discussions and provide valuable insights to both academic researchers and industrial practitioners as they explore the realm of flexible human–robot production.

3.1 Basic notions and connotation

In today's shift toward human-centric, sustainable, and resilient production, driven by Industry 5.0 (Xu et al., 2021b; Leng et al., 2022), industrial firms are striving to achieve three crucial objectives: 1) enabling transformable production processes that seamlessly accommodate new products without prolonged changeover times (Pedersen et al., 2016); 2) realizing flexible

Proactive Human–Robot Collaboration Toward Human-Centric Smart Manufacturing
https://doi.org/10.1016/B978-0-44-313943-7.00010-7

production of intricate and precise mechanical components and reducing dependence on manual labor (Cherubini et al., 2016); and 3) prioritizing the occupational health of employees by mitigating musculoskeletal disorders resulting from awkward postures, excessive exertion, and repetitive motions (Realyvásquez-Vargas et al., 2019). To address these imperatives, HRC has emerged as a prominent production paradigm, harmonizing the high precision and strength of robots with the advanced cognitive capabilities and adaptability of humans (Wang et al., 2020a). HRC systems can transition manufacturing processes toward flexible automation, optimizing productivity (Wang et al., 2020b; Wang and Wang, 2021).

In manufacturing, HRC allows humans and robots to work closely together (Wang et al., 2020a; Magrini et al., 2020). Over the past decade, numerous studies have explored HRC applications across various production activities. Research in human safety (Xu et al., 2021a), operator assistance (Zhang et al., 2020), and robot adaptive control (Garcia et al., 2019) has propelled the adoption of HRC in manufacturing tasks such as assembly, material handling, welding, and picking-and-placing. However, contemporary HRC architectures often remain confined to a master/slave mode. These systems struggle to learn and adapt to on-site conditions, lacking the ability to harness intellectual capabilities akin to humans.

In response to these limitations, a forward-looking paradigm within HCSM has emerged known as Proactive Human–Robot Collaboration (Proactive HRC) (Li et al., 2023). Proactive HRC is defined as "*a knowledge-driven, intelligent collaboration between human operators and robots, rooted in a comprehensive understanding of the on-site environment, transparent decision-making for resource allocation, and adaptable task execution in the face of industrial uncertainties.*" Within this framework, humans and robots collaboratively work proactively toward shared long-term objectives across various manufacturing activities and hierarchical tasks. Proactive HRC harnesses human-like intellectual capabilities and optimizes the synergy between human and robot complementary skills, resulting in heightened overall productivity and enhanced product quality.

3.1.1 Human operator engagement

Proactive HRC places a central emphasis on enhancing the human's physical, sensorial, and cognitive capabilities through robot automation. In this way, the human embodies the role of the "Operator 4.0" (Romero et al., 2016), a skilled collaborator who carries out tasks without experiencing physical or mental stress. This requires two fundamental prerequisites: ensuring human safety during collaboration and facilitating the intuitive perception of shared resources and services.

Human safety takes precedence in Proactive HRC connotation and remains a paramount concern. Established standards such as ISO/TR 7250 (International Organization for Standardization, 2010b), ISO 12100 (International Organization for Standardization, 2010a), and ISO/TR 14121-2 (International Organization for Standardization, 2012) outline specific requirements for ensuring human safety in manufacturing tasks, covering the prevention of physical injuries and the mitigation of risks related to occupational health. For instance, Schmidt and Wang (2014) implemented a system that monitored the minimum distance between humans and robots using 3D point cloud data from shop floors. This system could detect potential collision events, issuing warnings to operators, halting robot movements, and adjusting robot paths to reduce risk. Additionally, Peternel et al. (2019) explored methods to enhance comfort and safety by monitoring muscle fatigue in humans, issuing warnings for nonergonomic movements. This risk assessment approach minimizes hazards to human body parts.

The role of human intuition in HRC represents the connotation of the human natural ability to perceive shared information and make cognitive decisions necessary for effective teamwork. Multimodal communication technologies, including AR, voice recognition, haptic feedback, and Industrial Internet of Things (IIoT), enable humans to seamlessly perceive their surroundings. For example, Hietanen et al. (2020) developed an interactive AR system that empowered human operators to access real-time information about robot states and safety zone changes within the workspace. Liu et al. (2020a) introduced a human-centered robot control system that employed intuitive multimodal commands such as haptics, gestures, and voice. These enhancements in perceptual abilities cater to the interactive and cognitive needs of human operators, allowing them to observe and manage shared resources and services throughout the collaborative execution process.

3.1.2 Robot involvement and control

Collaborative robots within the Proactive HRC framework require timely adjustments and the ability to plan new manipulation motions, particularly when dealing with the production of high-mix, low-volume products. The programming of collaborative robots predominantly involves supervisory control and adaptive path planning.

The concept of supervisory control (Sheridan, 1986) enables robots to execute manufacturing tasks under human supervision while accommodating flexible human decisions. By utilizing continuous sensory feedback, such as force and impedance data, and responding to human commands, robots can dynamically adjust their stiffness and motion accuracy to align with human preferences. Supervisory robot control has been integrated into various Proactive HRC scenarios, encompassing physical human–robot interaction, teleoperation, robot path correction, and multimodal control, among others. Standard ISO 10218-2 (International Organization for Standardization, 2011) outlines power and force limitations and outlines hazardous situations involving collaborative robots in direct physical contact with human operators. For example, Kana et al. (2021) integrated impedance control and haptic interaction for human–robot comanipulation, enabling the robot to react to external forces exerted by humans through viscoelastic coupling. This supervisory control strategy combines the precision and strength of robots with the cognitive abilities of humans to a certain extent.

For adaptive path planning, robots can autonomously plan and execute motion trajectories, including end-effector positions, by leveraging inverse kinematics, dynamics systems, and embedded intelligent algorithms. These embedded algorithms enable robots to make decisions based on a holistic understanding of Proactive HRC scenarios, including human behavior, detected mechanical components, and robot status. Inverse kinematics and dynamics systems ensure that the robot achieves its target positions. Adaptive robot path planning holds significant importance in various Proactive HRC applications, such as collision avoidance, robot path re-planning, mobile robot assistance, and more. The standard ISO/TS 15066 (International Organization for Standardization, 2016) defines speed and separation monitoring criteria for robot adaptive motion to mitigate potential contact hazards. For instance, Ong et al. (2020) conducted experiments involving robot motion planning using the kinematics and dynamics library in the Robot Operating System (ROS) framework. By deducing semantic knowledge regarding human intentions and task structures, the robot could generate adaptive path planning to assist human operations and execute precise motions to complete tasks.

3.2 System architecture

The system architecture of Proactive HRC comprises three distinct layers: the task layer, the intelligence layer, and the application layer, as shown in Fig. 3.1.

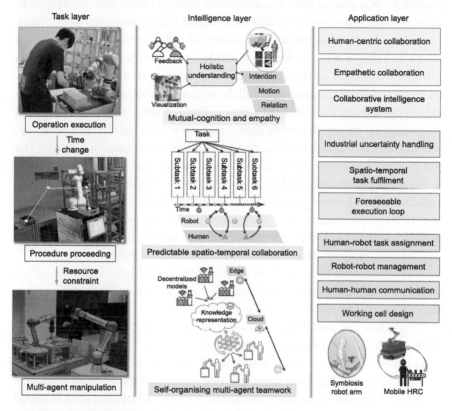

Figure 3.1 The system architecture of Proactive HRC.

The task layer encompasses a wide range of on-site situations encountered in diverse manufacturing tasks that involve both human and robot contributions. It involves three primary considerations. First, there is the collaborative execution of manufacturing tasks by both human and robot entities. Second, there is the dynamic reconfiguration of procedures as the task evolves over time. Third, the potential impacts of resource constraints and alterations are taken into account. Effective collaboration among multiple humans and robots is essential for handling complex tasks efficiently.

The intelligence layer aligns with the three considerations from the task layer and is the core to three overarching teamwork skills: 1) mutual-cognition and empathy, 2) predictable spatio-temporal collaboration, and 3) self-organizing multi-agent teamwork. These intellectual capabilities serve as incremental objectives within the Proactive HRC framework. During the operational execution phase, the mutual cognition and empathy module plays a pivotal role. It can deduce human intentions, comprehend robot motion states, and establish meaningful connections between them, achieving a comprehensive understanding of HRC scenarios. This enables robots to receive relevant feedback while providing humans with an intuitive visualization of the task execution process. As the task progresses, the predictable spatio-temporal collaboration module comes into play. It effectively decomposes tasks over time, allowing both humans and robots to familiarize themselves with their respective roles within the timeline. This ensures fluent task fulfillment and efficient coordination between the two parties. Finally, as more resources are introduced, the self-organizing multi-agent teamwork module takes charge. It distills knowledge representations from human guidelines, robot conditions, and task structures to optimize resource allocation. This knowledge representation can be encapsulated into cloud-edge computing resources and shared across multiple teams, promoting efficiency and collaboration across various operations.

The application layer serves as a demonstration of how Proactive HRC systems put the three intellectual modules into practice. These systems exhibit considerable potential in two main scenarios: symbiotic robot arms, which include augmented robots such as wearable exoskeletons, and mobile HRC, facilitating collaborative manufacturing activities in vast, unstructured spaces. The performance of these Proactive HRC systems can be significantly enhanced from three key perspectives. First, mutual-cognitive intelligence enables human-centric collaboration, addressing a wide range of human needs across various tasks. Empathetic collaboration ensures seamless teamwork with proactive human–robot involvement, promoting ease and comfort during joint operations. Collaborative intelligence systems maximize the utilization of human-qualified abilities and robot-suited roles, facilitating efficient task execution by leveraging their complementary strengths. Second, predictable intelligence equips a Proactive HRC system to effectively manage various uncertainties that may arise during task progression, ensuring a safe and predictable collaboration process. A spatio-temporal task fulfillment strategy can be devised, allowing both humans and robots to prepare for their respective roles in advance. A foreseeable execution loop enables both parties to anticipate each other's forthcoming actions, facilitating smooth coordination. Finally, with self-organizing intelligence, Proactive HRC systems can achieve hybrid human–robot task allocation, efficient robot-robot management, effective human–human communication, and well-designed work cell layouts. This elevates overall system efficiency and effectiveness.

The intellectual capabilities of Proactive HRC systems hold immense potential to revolutionize collaborative manufacturing processes and optimize task performance across various scenarios, as elaborated further in the following key characteristics.

3.3 Key characteristics

Advanced Proactive HRC is distinguished by three key characteristics: 1) mutual-cognition and empathy within the human–robot–workspace execution loop; 2) predictable spatio-temporal collaboration for efficient task fulfillment; and 3) self-organizing multi-agent teamwork with dynamic resource allocation.

3.3.1 Mutual-cognition and empathy

Mutual-cognitive capability in Proactive HRC involves *understanding dynamic human–robot relations within task structures and their operational intentions.* This understanding leads to the development of empathetic collaboration skills that facilitate ergonomic operations, aligning with teammates' preferences. The intelligence embedded in mutual cognition and empathy within HRC empower human operators with greater decision-making flexibility while enhancing robot manipulation skills for seamless teamwork.

Human operators excel in comprehending visible situations but may struggle to perceive information that is currently invisible, such as robot motion paths, updated operation procedures, and evolving manufacturing system conditions. Hence, a mutual-cognitive HRC system offers on-demand information support to humans, encompassing suggestions, guidance, warnings, and more. This knowledge support is derived from task processing status and procedural operation objectives at various task stages, providing human operators with a holistic grasp of task execution. Additionally, the human can convey new assessments of task progress to the robot through diverse multimodal communication methods, including Web interfaces, Digital Twins (DTs), VR, AR, haptic feedback, and brainwave interfaces. The mutual-cognitive service enhances human cognition and compensates for limitations in perceiving digital information.

Conversely, robots exhibit empathetic skills that go beyond adaptive path planning to enhance human well-being in HRC scenarios. Through knowledge acquisition regarding human intentions and ergonomic analysis of human actions, mutual-cognitive HRC systems allocate task operational strategies that cater to human needs. During task execution, the robot employs manipulation techniques that ensure comfort for human interaction, minimizing the impact of awkward postures and fatigue on the human's physical well-being. Moreover, the empathetic robot skill alleviates human mental stress by responding to changes in a worker's psychological state, offering clear and easily interpretable cues during noncontacting path motions.

As depicted in Fig. 3.2, the mutual-cognitive and empathetic capabilities within HRC systems rely on a human–robot–workspace perceptual loop, mutual-cognitive and empathetic decision-making processes, and cognitive services.

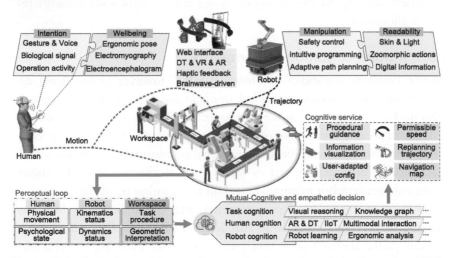

Figure 3.2 Mutual-cognitive and empathetic teamwork in Proactive HRC. (Adapted from (Li et al., 2023).)

3.3.1.1 Human–robot–workspace perceptual loop

For Proactive HRC systems, the human–robot–workspace perceptual loop stands as a critical prerequisite, enabling mutual cognition generation and on-demand collaborative behaviors. This loop contains human operator perception, readable robot status, and workspace parsing.

Human operator perception involves monitoring a worker's physical and psychological states during co-working, with the aim of comprehending human operators' operational intentions and well-being in HRC tasks. Human intention (Abdelkawy et al., 2020; Zhang et al., 2019) includes long-term goal planning for collaborative task completion with robots, such as responses to stepwise operations and unexpected events. Human well-being (Buerkle et al., 2022), on the other hand, pertains to the expectations regarding resource allocation and task completion in teamwork. This is reflected in physical experiences like fatigue and mental satisfaction. For example, overloads and improper postures may lead to health risks and occupational injuries. Human operator perception enables robot teammates to adapt effectively, particularly in the following Proactive HRC implementation aspects: 1) adaptive robot control with awareness of human intentions, generating robot motion planning aligned with human actions; 2) supervisory robot control incorporating human cognition, allowing human operators to guide robot execution using natural action commands; and 3) empathetic robot skills for human well-being, where robots perform assist-as-needed movements and ergonomic path motions to enhance human comfort.

Readable robot status aims to intuitively convey robots' actions and planning to human operators. This feedback enhances human cognitive decision-making. Past efforts in robot status understanding include nonverbal indicators such as tower lights (Tang et al., 2019a), skin (Pang et al., 2020), and zoomorphic actions (Sauer et al., 2021) (eye gaze, head orientation, and arm movement), as well as digital approaches like DT, VR, and AR. These shared robot state variables can be adjusted and corrected through human intervention or continuous self-adjustment. Instead of storing robot status invisibly in controllers, Proactive HRC systems encourage legible robot motions, facilitating humans' ability to interpret robot intentions. Readable robot status benefits in the following ways: 1) promoting humans' prompt decision-making and reactions to robot legibility signals during manufacturing operations; 2) facilitating easy and intuitive teamwork in industrial settings with one or more robots requiring varying levels of priority; and 3) enabling timely corrections for robot infeasible status to maintain efficient robot behavior.

Workspace parsing aims to interpret geometric and semantic knowledge within the working environment, including the detection of static and dynamic objects (Rosenberger et al., 2020), spatial pose estimation (Adeli et al., 2020), scene construction (Moon and Lee, 2018), and task interpretation (Mateus et al., 2020). As manufacturing tasks progress, workspace parsing offers real-time information to support adaptive robot control and intuitive human operations. This area explores portable and flexible visual sensor systems such as binocular cameras, depth cameras, and LiDAR, complemented by advanced computer vision and Deep Learning (DL) techniques. Workspace parsing knowledge allows both humans and robots to maximize their active roles in co-working: 1) enabling on-demand task adjustments in response to changes in HRC settings, such as time-varying subtask operations and predictable collision avoidance; 2) facilitating high-precision and dexterous robot operations by providing refined spatial and location information of targeted parts; and 3) empowering active human decision-making and robot planning based on holistic scene construction and an understanding of task processes.

3.3.1.2 Mutual-cognitive and empathetic decision

Based on the perceptual results, Proactive HRC systems can achieve empathetic task-planning decisions to facilitate bi-directional collaboration and ergonomic teamwork. The decision intelligence is delineated through three key components: 1) task cognition within operational sequences; 2) augmented human cognition with advanced operational skills; and 3) robot cognition aimed at operating human-desired manipulations.

Task cognition in Proactive HRC revolves around the comprehension of human–robot–task structures and their associated intentions (Ahn et al., 2018), extracted from perceptual data generated during task processes. This intelligence is then embedded into the decision-making process to plan mutually beneficial ongoing operations. In Proactive HRC, task cognition empowers both humans and robots with dynamically evolving operation plans to meet the demands of specific tasks. These cognitive co-working planners, integrated into Proactive HRC systems, effectively optimize the synergy between human intuition and robot adaptability. They are developed by cognitive planners into the execution loop, delivering mutual support and intuitive guidance for appropriate operations through advanced technologies like AR, DT, and IIoT. Furthermore, task cognition, by understanding the intricate human–robot–task relationships, allocates tasks to human and robot agents while maintaining alignment with the overarching collaboration objectives. Ultimately, these cognitive task planners enable the ongoing refinement of task progress by optimizing resource allocation.

Enhancing human cognition within HRC systems is essential to elevate workers' perception capabilities (Arkouli et al., 2021), manual operational skills (Chen et al., 2020), and communication proficiency with robots (Liu et al., 2020b). The enhanced human cognition equips operators with superior operational abilities in Proactive HRC, particularly in three key dimensions: 1) self-learning capacity for personal growth and self-actualization through continuous knowledge acquisition supported by domain-specific knowledge within the execution loop; 2) augmented decision-making skills, empowered by access to essential procedural information, enabling operators to anticipate future events and make informed decisions based on a holistic understanding of HRC tasks; and 3) a more collaborative and user-friendly working experience, where individual well-being is improved through flexible robot control that adapts to interactions and responds to further task decisions, allowing operators to fine-tune or re-plan robot motions as needed.

Robot cognition aims to imbue robots with the capacity to comprehend human behaviors and intentions, align with common task objectives, and plan adaptive manipulations in accordance with human preferences. In Proactive HRC, a collaborative robot with cognition transcends basic attributes such as collision avoidance to evolve into an intelligent agent endowed with empathetic teamwork skills (Costantini et al., 2019). Key aspects of this evolution include: 1) human-like cognition, achieved through knowledge representation learning of human intentions and the development of task planning strategies to facilitate adaptive, human-desired manipulations; 2) robot learning, enabling robots to actively assist human operators and rapidly devise suitable motions by imitating and learning from human operations and commands; and 3) empathic teamwork skills, involving the analysis of ergonomic satisfaction with human postures during collaboration. Robots dynamically adjust end-effector positions and orientations to ensure comfortable human engagement, enabling operations with forces, poses, and speeds that align with human preferences and mitigate physical and mental stress.

3.3.1.3 Cognitive service

Cognitive services within Proactive HRC systems are centered on facilitating the transmission of mutual-cognitive task planning to both human and robotic agents through natural and nonmisleading communication channels. These services are seamlessly integrated into the collaborative workflow of HRC systems, encompassing activities such as essential information sharing and on-demand guidance and control commands.

The provision of cognitive services hinges on establishing multimodal and bi-directional information exchange between humans and robots. This exchange occurs through diverse means, including web interfaces displayed on screens and tablets (Rey et al., 2021), the recognition of gestures and motion captured by camera devices, speech information relayed via microphones and speakers, haptic sensors (Kahanowich and Sintov, 2021), DT, VR, AR, and even the analysis of brainwave data (Liu et al., 2022). To illustrate the practicality of cognitive services, several use cases have been explored, such as welding, assembly, co-grasping, and robot route planning within HRC systems.

To ensure the natural and intuitive delivery of cognitive services in Proactive HRC, an MR approach that incorporates DT and AR technologies has emerged as a promising avenue for further exploration. The MR-based execution loop, which combines the capabilities of DT and AR, enhances human context awareness by offering procedural guidance, visualizing critical information, and configuring user-specific settings in a fusion of virtual and real environments. Within this MR environment, robot motions are simulated and predicted in real-time, accounting for factors like permissible speed (Byner et al., 2019), trajectory re-planning (Wang et al., 2017), and navigation mapping (Wang et al., 2021), thereby minimizing the occurrence of runtime errors in robotic operations. Additionally, the MR-based cognitive services facilitate natural communication across multiple human and mobile robot agents operating within expansive spatial settings. This inclusive environment accommodates interactions among one or more agents, allowing for the exchange of information and communication with various parties, each possessing distinct levels of priority and roles in the collaborative workflow.

3.3.2 Predictable spatio-temporal collaboration

Predictable spatio-temporal collaboration within the Proactive HRC framework serves the goal of *generating optimal task fulfillment plans and forecasting the execution processes over time*. This is achieved by leveraging predictive insights into human–robot motions and the allocation of system resources, i.e., a holistic perception of the manufacturing activities.

In the context of a production process, a multitude of manufacturing knowledge, including resource management and operational objectives, should be considered. For instance, the overall manufacturing process contains hierarchical subtasks or stages with evolving task progress (Johannsmeier and Haddadin, 2016). Within one specific stage, it is hard for a human operator to comprehend the procedural details of the entire manufacturing task and plan rational strategies for upcoming operation goals, especially in the face of unexpected events. Predictable spatio-temporal collaboration addresses this by learning knowledge from predefined task schedules and inferring the current status of task progression. By aligning real-time operational data with learned manufacturing knowledge representation, the system generates task planning decisions for the ensuing stage of the task. These decisions are assigned to both human and robotic agents to ensure the fluent completion of the task, particularly in scenarios characterized by industrial uncertainties.

In specific instances, executing decisions in a task stage may introduce psychological stress for a human operator who is uncertain about robot motions and the unfolding future events. To tackle this, the predictable spatio-temporal collaboration models and predicts human movements within the execution loop. It goes further by providing human operators with a preview of robot trajectories (Kawasaki and Seki, 2021), allowing robots to proactively plan their motions in advance. This predictive intelligence reduces the psychological workload on humans and promotes a more foreseeable accomplishment of HRC tasks, all while encouraging natural human participation and proactive robot behavior.

The realization of predictable spatio-temporal collaboration in Proactive HRC is achieved through a multifaceted approach, involving the analysis of human uncertainty and error-prone operations, a task precedence constraint planner, spatio-temporal task fulfillment, and the establishment of a foreseeable execution loop. These elements work in concert to bring predictability and efficiency to the collaborative manufacturing process, as illustrated in Fig. 3.3.

3.3.2.1 Human uncertainty and error operation

Human operations in HRC systems are characterized by uncertainty, such as uncertain movements and incorrect actions. These human behaviors may disrupt corresponding robot manipulations and, in more severe cases, pose safety risks (Bi et al., 2021). To ensure the fluent progression of tasks and to counterbalance the unexpected disturbances introduced by human factors, it is imperative to undertake an analysis of human uncertainty and error-prone operations. This analysis serves as a foundational prerequisite for achieving predictable spatio-temporal collaboration within the Proactive HRC paradigm.

The identification and prediction of human uncertainty and error operations can be accomplished through a diverse array of methodologies and sensor technologies. These include but are not limited to torque sensors, laser scanners, visual cameras, and DT modeling. By systematically

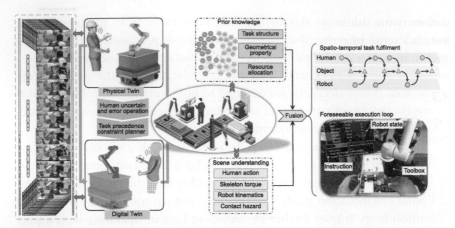

Figure 3.3 Predictable spatio-temporal collaboration in Proactive HRC. (Adapted from (Li et al., 2023).)

scrutinizing human uncertainty, a Proactive HRC system can optimize its operational efficiency and safety measures. This may involve mitigating contact hazards, detecting instances of geometric occlusion, fine-tuning robot kinematic configurations, estimating human joint torque, and evaluating ergonomic considerations. Consequently, the robot can perform tasks at speeds that align with human-permissible rates, engage in risk-based replanning when necessary, and minimize waiting and idle time (Unhelkar et al., 2018).

In Proactive HRC, the concept of a digital human is a pivotal component for simulating and predicting a worker's physical actions and psychological states, as illustrated in the left section of Fig. 3.3. This digital representation is continuously updated and in sync with the physical human worker. Through this integration, the system gains the capability to anticipate a worker's future focus, behaviors, and assess their physical and mental workload within digital environments. For instance, the digital system can simulate how often an individual worker gazes at a robot during different operations and proactively furnish them with previewable robot trajectories at precisely the right moment. Furthermore, a digital human model enriched with knowledge in production management assumes a prominent role, enabling it to discern human uncertainty and correct erroneous operations.

The prediction of human uncertainty and incorrect operations equips HRC systems with the ability to swiftly realign irregular or hazardous situations with the intended co–working strategy. This proactive approach fosters a sense of compassion and coevolution between human and robot agents, in a safer collaborative environment.

3.3.2.2 Task precedence constraint planner

Operation precedence constraints play a pivotal role in hybrid HRC task execution. Consider a scenario where certain manufacturing stages necessitate manual human operations, which are subsequently followed by robot manipulations or handovers. This gives rise to a task precedence constraint planner, an integral component of prior production knowledge for the realization of predictable spatio–temporal collaboration in HRC systems.

The representation of task precedence constraints can be acquired through an array of methods, such as probabilistic models (Pulikottil et al., 2021), tree structures (Tang et al., 2019b), Knowledge Graph (KG) (Ding et al., 2019), Reinforcement Learning (RL) (Yu et al., 2021), and more. Understanding specific task constraints can facilitate both real-time scheduling of long-term task fulfillment and generate strategies for close-proximity execution. To illustrate, consider product-specific structural constraints that serve as a foundation for dynamic task strategy adjustments when HRC systems are engaged in tasks like assembly or disassembly. In a different context, comprehending constraints within machining processes ensures the high surface quality and precision in HRC grinding systems.

To infer knowledge of task precedence constraints regarding product variants, cutting-edge technologies, including KG and RL, offer viable solutions. The comprehensive task structure involved in the manufacturing process of a product can be acquired and integrated within a KG. The manufacturing constraints pertaining to individual product components can be represented as subgraphs within the overarching task structure. These subgraphs serve as representations of different stages in HRC tasks, whereas the edges within the KG depict the sequential operation sequences within specific subtasks. Moreover, the KG possesses the capability to transfer existing knowledge and infer general rules regarding various node attributes, thereby enabling the learning of task precedence constraints for product families that exhibit similar structures. In the context of HRC production for new products with extensive variations, RL methods can be effectively employed to acquire the task structure through iterative simulation and optimization of working processes within DT environments.

3.3.2.3 Spatio-temporal task fulfillment

Spatio-temporal task fulfillment is a critical aspect for the generation of comprehensive operation plans for hierarchical HRC subtasks in forthcoming stages. This process relies on current scene comprehension and prior manufacturing knowledge. Spatio-temporal task fulfillment can be categorized into two primary scenarios (as depicted in the upper-right section of Fig. 3.3). The first scenario pertains to long-term HRC workflows, achieved through the learning of task precedence constraints, which serve as pivotal prior knowledge. Concurrently, real-time monitoring and logical reasoning of the current scene understanding provide immediate feedback, ensuring the fluent completion of HRC tasks. The second scenario addresses situations characterized by human uncertainty and erroneous operations within the current scene (Askarpour et al., 2019). In such cases, the HRC system necessitates dynamic re-planning and adjustment of forthcoming task operations, leveraging prior manufacturing knowledge to facilitate effective teamwork execution.

Advanced techniques, including KG, RL, and DT, play a key role in realizing spatio-temporal task fulfillment. Currently, the emphasis within HRC task fulfillment predominantly revolves around resolving task scheduling challenges, with less attention directed toward the organization of sequential operation sequences for both humans and robots in future stages. This hinders their ability to truly collaborate as tasks progress. To enable the widespread implementation of spatio-temporal task fulfillment in Proactive HRC, several feasible solutions can be considered.

One approach involves constructing a KG of working processes, including various hierarchical levels such as components, operations, and task decomposition. This HRC KG effectively links diverse forms of prior knowledge, including task structure, geometrical attributes, and resource allocation, facilitating task planning in each stage with precise operational instructions.

Another approach integrates prior knowledge into RL algorithms for HRC task planning. In this scenario, when generating task planning directives for a new and previously unseen task, RL algorithms expedite the search for efficient strategies by circumventing redundant trial-and-error processes.

The final approach involves the establishment of knowledge-based DT environments for HRC task re-planning. These knowledge-based DT environments are adept at distilling the current scene's representation, encompassing elements such as human actions, skeletal torque, robot kinematics, and potential contact hazards. By seamlessly incorporating scene understanding into the overarching prior knowledge of a given task, knowledge-based DT systems dynamically simulate, predict, and re-plan specific next steps for both human and robotic agents in response to changing HRC settings.

3.3.2.4 Foreseeable execution loop

The foreseeable execution loop plays a pivotal role in facilitating proactive robot actions that align with human intentions in advance, avoiding potential error operations. This proactive approach is exemplified in the lower-right section of Fig. 3.3. On one side, the robot utilize predictive algorithms to anticipate, re-plan, and reach interaction points ahead of time, predicated on the forecasted future movements of humans. On the other side, the HRC system exercises its capability to make suitable decisions and assign the robot's subsequent operations through early detection of human motions and operational intentions.

In a foreseeable execution loop, the utilization of AI models, including DL, TL, and multimodal learning, plays a critical role in enhancing the accuracy and robustness of human action prediction. In the context of Proactive HRC, further enhancements can be achieved by incorporating a knowledge model into the loop. This knowledge model learns and assigns priority to different prediction elements based on their relevance. For instance, human actions involving turn-taking events (Zhou and Wachs, 2018) and potential risk factors demand heightened attention, such as actions involving handovers, interactions, and potential contact hazards. Conversely, smooth and sustained human operations that do not exhibit significant time constraints, such as static actions like standing in one location during the unscrewing process, may receive relatively lower priority.

Furthermore, the foreseeable execution loop should be intimately connected with the DT environment. This integration enables real-time simulation and prediction of various factors, including human fatigue levels, ergonomic working conditions, and the potential for runaway motions. With both the DT and knowledge model integrated into the loop, the robot can infer, reason, and proactively plan suitable path trajectories in advance. This proactive approach reduces human waiting time, enhances overall efficiency, and fosters a more natural and seamless collaborative environment for human–robot task execution.

3.3.3 Self-organizing multi-agent teamwork

Self-organizing multi-agent teamwork within Proactive HRC centers on *the coordination of numerous human and robotic agents across expansive and unstructured spaces*. This coordination is predicated on their distinct qualifications and operational roles, with the aim of meeting stringent requirements such as execution time, resource availability, and minimal energy consumption. The ensemble of agents in a Proactive HRC system includes human operators, robot arms, and mobile robots. Serving as the central hub of intelligence in Proactive HRC, self-organizing capabilities facilitate seamless information exchange among multiple agents, enabling globally optimized collaborative processes. The acquisition of self-organizing knowledge is underpinned by four key facets: 1) task structures and decomposition, i.e., understanding the intricate architecture and division of tasks; 2) dynamic environment and event relations, which grasp the ever-changing environmental conditions and event interdependencies; 3) resource occupation constraints and ergonomics, which incorporate considerations related to resource usage and ergonomic factors; and 4) capable agent execution rules in unstructured space, which learn the operational rules for proficient agents functioning in unstructured environments.

Illustrated in Fig. 3.4 is a demonstration of self-organizing multi-agent teamwork in Proactive HRC. At the organizational level, the design of a working cell for multiple agents is a crucial prerequisite. This design distills the knowledge representation of general rules of human–robot operations. Then, the HRC system can swiftly leverage this acquired knowledge to generate task arrangements, even when confronted with product variations.

On a task-oriented level, self-organizing multi-agent teamwork hinges on three pivotal aspects: 1) information communication between multiple humans, which facilitates seamless information exchange among diverse scenarios and different locations; 2) resource allocation and management among collective robots, which manages resources to optimize task planning; and 3) interaction and role-playing among hybrid human–robot agents, which coordinate and assigning global tasks among the various agents.

Techniques for information exchange, task planning, and task assignment facilitate the interplay between hybrid human and robotic agents, enabling them to leverage their unique capabilities for specific tasks within the Proactive HRC framework (Zhang et al., 2021).

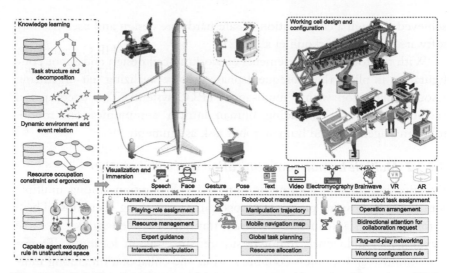

Figure 3.4 Self-organizing multi-agent co-working in Proactive HRC. (Adapted from (Li et al., 2023).)

3.3.3.1 Working cell design and configuration

The working cell design and configuration lies in crafting adaptable workstations and establishing a layout for human and robotic participants. This phase is instrumental in realizing self-organizing multi-agent teamwork within the Proactive HRC framework.

When it comes to crafting configurable workstation layouts that cater to task-specific robots and human operators, there remain two promising research directions deserving further exploration. The first one involves AI-driven HRC cell generation, leveraging methodologies such as RL. AI-based HRC cell generation has the capacity to learn knowledge from historical experience related to workstation design. It can intelligently devise human–robot layouts through iterative optimization while adhering to essential constraint rules. For instance, it ensures that humans are not tasked with physically demanding, repetitive labor that exceeds their physical capabilities (Mauri et al., 2019). By striking this balance, the HRC cell optimizes work distribution to prevent undue strain on human operators.

The second direction centers on wearable and augmented robots, particularly lightweight exoskeletons equipped with either active or passive actuation capabilities. These wearable exoskeleton robots serve to augment human payload capabilities while fostering symbiotic relations. This can be achieved through a combination of safe hardware design and user-friendly software interfaces (Arkouli et al., 2021).

With the foundational elements of working cell design and robot configuration in place, the subsequent phase involves realizing self-organizing resource allocation within multi-agent HRC systems. This encompasses various dimensions, including human–human communication, robot–robot management, and human–robot task assignment.

3.3.3.2 Multiple human communication and collaboration

Effective information communication and collaboration among multiple human operators are prevalent across various working scenarios. These communication and interactions may transpire between parties in close proximity or individuals remotely situated in different locations. To ensure intuitive and seamless communication and collaboration among multiple human participants, it is essential to adapt diverse user roles, operational behaviors, and interactive tendencies. In complex industrial settings, human–human communication and collaboration are often characterized by considerations related to social well-being and esteem needs.

In Proactive HRC, multiple human collaboration contains role assignments, human resource management (Shang et al., 2018), expert guidance (Wang et al., 2019), and interactive co-manipulation (Tokadlı and Dorneich, 2019). The integration of AR and DT technologies introduces communication tools that facilitate self-organizing collaboration among multiple human participants. To achieve this, three critical aspects deserve particular attention.

The first aspect should ensure time consistency in multiple human communication. When various interaction events occur, it is imperative to guarantee synchronous access without conflicts in network communication mechanisms. Consequently, different working scenarios can transition seamlessly and be simultaneously delivered to multiple individuals via display devices, facilitating information exchange and collaborative experience.

The second aspect is spatial consistency for scenario visualization among multiple individuals located in remote spaces. Leveraging tracing and positioning technologies like Simultaneous Localization and Mapping (SLAM), human operators situated in diverse locations can collectively access a holographic display presenting the same view of a manipulated product. This shared visual perspective enables them to engage in collaborative operations toward a common task objective.

Lastly, emphasizing visible fidelity and immersion during information communication is crucial for enhancing the naturalness and fluency of interactions among multiple human teammates. For example, when an on-site worker receives domain knowledge from a remote expert, AR and DT tools can optimize their coexistence by adapting to individual action habits, personalized interactive gestures, and designated roles, thereby elevating the overall user experience.

3.3.3.3 Multiple robot management and task planning

Multiple robot management aims for optimizing task planning and resource allocation among a variety of robotic entities, including fixed robot arms, Automated Guided Vehicles (AGVs), and mobile robots. Within industrial contexts, effective collaboration among multiple robots can yield benefits such as optimal manipulation trajectories (Hassan et al., 2015), well-structured mobile navigation maps (Chang et al., 2020), and efficient global task distribution (Liau and Ryu, 2022). This shift toward multiple robot management and task planning allow HRC systems, to transition them from conventional leader–follower models to intelligent multi-agent systems characterized by improved fault tolerance.

In Proactive HRC, the self-organizing teamwork among multiple robots assumes a pivotal role in minimizing task completion duration, optimizing the utilization of capable agents, and mitigating ergonomic risks to human collaborators. These objectives can be achieved by considering several key aspects:

a) (Task Decomposition) Employ cutting-edge techniques, including KG, to deconstruct complex manufacturing tasks into a sequence of executable subtasks. These subtasks are represented as nodes within a graph structure spanning various task levels, complete with attributes denoting resource constraints, execution timeframes, relation rules, and associated costs.

b) (Agent Assignment Preference) Leverage the knowledge distilled from task decomposition to analyze the characteristics and capabilities of different robot agents. This analysis facilitates the determination of each agent's preferred assignments, including manipulation trajectories, navigation paths, and subtasks. By aligning agent capabilities with task requirements, optimal resource allocation can be achieved.

c) (Autonomous Exploration) Empower each robot within the Proactive HRC system with autonomous exploration capabilities. This enables robots to dynamically interact with their environment, assess relationships among surrounding objects, and engage in communication with other robotic entities to receive feedback pertaining to task execution goals.

3.3.3.4 Hybrid multi-agent interaction and task assignment

Hybrid multi-agent interaction and task assignment are essential components of achieving effective communication and coordination between human operators and diverse robotic systems. The primary goal is to determine their dynamic roles within a collaborative workflow and optimize global task arrangements to facilitate seamless teamwork. Hybrid multi-agent HRC emerges as an important concern, particularly suited for the production of small to medium-volume products with significant variations, often encountered in unstructured spaces such as aircraft manufacturing.

In the context of hybrid multi-agent HRC, different types of agents, including industrial robot arms, high-payload mobile robots, collaborative robots, and human operators, play distinct yet complementary roles:

a) (Industrial Robot Arms) These systems excel in executing ultraprecise operations like welding and machining processes.

b) (Mobile High-Payload Robots) Designed for tasks requiring heavy manipulation and substantial workloads in large spaces, such as material handling.

c) (Collaborative Robots) Assist human operators in performing dexterous tasks, reducing their workload.

To realize self-organizing hybrid multi-agent teamwork in Proactive HRC, further research explorations are necessary, focusing on key aspects:

a) (Bi-directional Collaboration) Develop mechanisms for bi-directional collaboration considering resource coordination between human and robotic agents (Patnayak et al., 2021). KG methods can model various human and robotic operations and establish their relations within a task, facilitating efficient collaboration in operations like product state monitoring, picking, placing, remote control, and screw mounting.

b) (Plug-and-Play Networking) Establish plug-and-play networking and connectivity solutions between hybrid agents in HRC systems. Leveraging IIoT techniques based on cloud-edge computing, these systems can provide multi-agent communication and extensible network access (Costantini et al., 2019). This approach ensures flexibility in configuring different human–robot teams and enables rapid creation of agent-based services.

c) (General Rules and Knowledge Sharing) Facilitate the acquisition and utilization of general rules of hybrid human–robot collaborative workflows (Mokhtarzadeh et al., 2020). The prior knowledge can be used to guide human operators and train robots in making collaborative decisions when integrating new heterogeneous robots into the team. By leveraging these established rules and knowledge representations, the system can ensure effective coordination among new human–robot teammates, like transfer skills of human operator guidance on what to do and robot learning on how to generate collaborative decisions.

3.4 Intelligent robot control and human assistant system

The integration of cognitive, predictable, and self-organizing intelligence within the control execution loop empowers intelligent robot control and a human assistant system, as illustrated in Fig. 3.5. Within this framework, knowledge flows from the bottom to the top, facilitating efficient collaboration between robots and human operators.

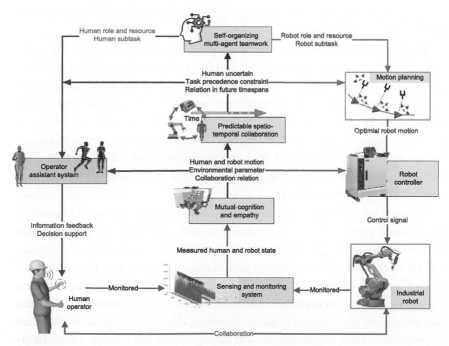

Figure 3.5 Intelligent robot control and human assistant system in Proactive HRC. (Adapted from (Li et al., 2021).)

In this process, the middle cognition module serves as a critical component, enabling the acquisition of high-level knowledge derived from spatio-temporal HRC tasks and translating it into semantic knowledge. As a result, collaborative intelligence is generated across different levels within the middle cognition module. With the collaborative knowledge, robots (as depicted on the left side) and human operators (as shown on the right-hand side) can work together proactively to execute manufacturing tasks, guided by specific instructions that flow from top to bottom.

3.4.1 From mutual-cognitive intelligence level

The holistic comprehension of the HRC scenario, encompassing real-time states on collaborative robots, human motions, and intentions, is facilitated by the human–robot–workspace perceptual loop. These states and information serve as feedback for the robot's motion controller, enabling proactive execution of manufacturing tasks, as depicted in the lower section of Fig. 3.5. Then, the mutual-cognitive intelligence between robots and humans can be embedded within their collaborative operations, such as haptic interactions at close range or proactive robot execution with intuitive human cooperation.

In scenarios involving close or direct physical contact, monitoring physical parameters of haptic interactions and interpreting the working environment relies on sensor systems. The robot controller, by sensing force and moment signals, can employ compliance control strategies like impedance control (Yu et al., 2019). This dynamic adjustment of contact forces and robot positions ensures safe interactions. For instance, it can prevent damage to objects being handled or harm to humans in HRC scenarios by involving collision-free motion planning, safety-rated monitored stop, and power/force limiting control.

In cases of long-range collaboration, an operator assistant system and multimodal programming and robot control can facilitate proactive bidirectional engagement between humans and robots. The operator assistant system tracks the robot's motions and plans, providing real-time information support for intuitive human task execution. Multimodal programming and robot control incorporate human gestures, speech, motion, intention, and other data from sensor systems, allowing the robot to proactively adapt its behavior. This adaptation ensure human operators work comfortably while reducing task completion times and increasing overall production efficiency.

3.4.2 From predictable intelligence level

Predicting the trajectories of both human and robot motions, as well as anticipating the human's next intentions and their operations in future subtasks, can be leveraged for both robot motion planning and the operator assistant system, as depicted in the central segment of Fig. 3.5. This predictive intelligence means foreseeable and proactive robot control, in a feedback loop.

Proactive motion planning allows the robot controller to proactively adjust its motions in advance, optimizing performance in various aspects. This includes tasks such as collision avoidance between the human and the robot, enhancing overall work efficiency.

As this predictive intelligence incorporates feedback information on human–robot operations across spatio-temporal domains, the operator assistant system aids the human in comprehending their subtask within the overall collaborative process, thereby mitigating the risk of operational errors (De Pace et al., 2020). Besides, spatio-temporal collaboration prediction also provides the appropriate cooperation dynamics in future subtasks. This information can be informed to human and robotic agents in advance for decision-making and effective task allocation.

3.4.3 From self-organizing intelligence level

The self-organizing intelligence for Proactive HRC plays a pivotal role in determining the roles of both human and robot parties involved and effectively allocating the necessary resources and subtasks to each participant. The outcomes yielded by these self-organizing decisions serve as the primary inputs for both robot motion planning and operator assistant systems, as illustrated in the upper section of Fig. 3.5. In detail, drawing from the information measured and predicted regarding the states of both humans and robots through mutual cognition and predictable collaboration, specific resources and subtasks are distributed to the human and robotic teammates. This allocation process adheres to specific criteria, such as optimizing task execution time or minimizing energy consumption (Mohammed et al., 2017).

In execution processes, the robot's motion is optimized to execute the designated subtask in alignment with generated policies. These policies considers human safety, resource availability, and the suitable operation time. Meanwhile, the operator assistant system retrieves information from its database in response to the assigned subtask. It provides valuable instructions and support to assist the human worker in effectively carrying out their designated role.

It is noted that despite some uncertainties existing due to the human operator's presence within the HRC loop, the self-organizing teamwork remains agile, capable of dynamically adjusting resource and task allocation in response to human behavior. This adaptability ensures the smooth and uninterrupted execution of the overall task.

3.5 Chapter summary

In the era of HCSM, both industry and academia are working towards achieving a practicable human–robot collaborative production environment, in which the Proactive HRC paradigm assumes a crucial significance. This chapter presents a comprehensive analysis of our perspective that Proactive HRC systems can encompass mutual-cognitive, predictable, and self-organizing intelligence to facilitate flexible, natural, and efficient production processes. We provide a systematic exposition of the emerging trend towards Proactive HRC, highlighting the growing demands and motivations behind it. Furthermore, we present our findings on the conceptual framework and referential architectures of this foreseeable manufacturing paradigm, while also covering the challenges it faces and the enabling technologies that can support its realization.

Central to our contribution is the establishment of the Proactive HRC architecture, offering a comprehensive definition and detailed exposition across four core facets: 1) mutual-cognition and empathy, 2) predictable spatio-temporal collaboration, 3) self-organizing multi-agent teamwork, and 4) intelligent robot control and human assistant systems. The first three modules represent continuously evolving intelligent capabilities within Proactive HRC, while the final one encompasses the enabling control technologies required for practical implementation. This theoretical foundation serves as a springboard for researchers dedicated to investigating theories and techniques underpinning mutual-cognitive capabilities within Proactive HRC, such as visual reasoning and KG methods. Recent studies have begun showing interest in predictable intelligence within Proactive HRC, particularly in the realm of motion trajectory prediction. Moreover, it is foreseeable that self-organizing capabilities for multi-agent teamwork in Proactive HRC will attract increasing attention in the future. Lastly, with the emergence of more algorithms for compliant robot control and proactive path planning, addressing the challenges associated with deploying Proactive HRC applications in modern factories still demand years of effort.

We hold a deep conviction that the Proactive HRC trend will show widespread deployment in modern enterprises, facilitating large-scale, flexible, and automated production of personalized and intricately designed products. The rapid development of technologies related to cognitive computing, knowledge acquisition, and automation theories is accelerating the transition to the Proactive HRC paradigm, reshaping the production landscape across industries. Our hope is that these early insights into Proactive HRC will ignite a wealth of insightful discussions, validations, and constructive debates within the research community.

References

Abdelkawy, H., Ayari, N., Chibani, A., Amirat, Y., Attal, F., 2020. Spatio-temporal convolutional networks and n-ary ontologies for human activity-aware robotic system. IEEE Robotics and Automation Letters 6, 620–627.

Adeli, V., Adeli, E., Reid, I., Niebles, J.C., Rezatofighi, H., 2020. Socially and contextually aware human motion and pose forecasting. IEEE Robotics and Automation Letters 5, 6033–6040.

Ahn, H., Choi, S., Kim, N., Cha, G., Oh, S., 2018. Interactive text2pickup networks for natural language-based human–robot collaboration. IEEE Robotics and Automation Letters 3, 3308–3315.

Arkouli, Z., Kokotinis, G., Michalos, G., Dimitropoulos, N., Makris, S., 2021. AI-enhanced cooperating robots for reconfigurable manufacturing of large parts. IFAC-PapersOnLine 54, 617–622.

Askarpour, M., Mandrioli, D., Rossi, M., Vicentini, F., 2019. Formal model of human erroneous behavior for safety analysis in collaborative robotics. Robotics and Computer-Integrated Manufacturing 57, 465–476.

Bi, Z., Luo, C., Miao, Z., Zhang, B., Zhang, W., Wang, L., 2021. Safety assurance mechanisms of collaborative robotic systems in manufacturing. Robotics and Computer-Integrated Manufacturing 67, 102022.

Buerkle, A., Matharu, H., Al-Yacoub, A., Lohse, N., Bamber, T., Ferreira, P., 2022. An adaptive human sensor framework for human–robot collaboration. The International Journal of Advanced Manufacturing Technology 119, 1233–1248.

Byner, C., Matthias, B., Ding, H., 2019. Dynamic speed and separation monitoring for collaborative robot applications—concepts and performance. Robotics and Computer-Integrated Manufacturing 58, 239–252.

Chang, W., Lizhen, W., Chao, Y., Zhichao, W., Han, L., Chao, Y., 2020. Coactive design of explainable agent-based task planning and deep reinforcement learning for human–UAVs teamwork. Chinese Journal of Aeronautics 33, 2930–2945.

Chen, C.S., Chen, S.K., Lai, C.C., Lin, C.T., 2020. Sequential motion primitives recognition of robotic arm task via human demonstration using hierarchical BiLSTM classifier. IEEE Robotics and Automation Letters 6, 502–509.

Cherubini, A., Passama, R., Crosnier, A., Lasnier, A., Fraisse, P., 2016. Collaborative manufacturing with physical human–robot interaction. Robotics and Computer-Integrated Manufacturing 40, 1–13.

Costantini, S., De Gasperis, G., Migliarini, P., 2019. Multi-agent system engineering for emphatic human–robot interaction. In: 2019 IEEE Second International Conference on Artificial Intelligence and Knowledge Engineering (AIKE). IEEE, pp. 36–42.

De Pace, F., Manuri, F., Sanna, A., Fornaro, C., 2020. A systematic review of augmented reality interfaces for collaborative industrial robots. Computers & Industrial Engineering 149, 106806.

Ding, Y., Xu, W., Liu, Z., Zhou, Z., Pham, D.T., 2019. Robotic task oriented knowledge graph for human–robot collaboration in disassembly. Procedia CIRP 83, 105–110.

Garcia, M.A.R., Rojas, R., Gualtieri, L., Rauch, E., Matt, D., 2019. A human-in-the-loop cyber-physical system for collaborative assembly in smart manufacturing. Procedia CIRP 81, 600–605.

Hassan, M., Liu, D., Paul, G., Huang, S., 2015. An approach to base placement for effective collaboration of multiple autonomous industrial robots. In: 2015 IEEE International Conference on Robotics and Automation (ICRA). IEEE, pp. 3286–3291.

Hietanen, A., Pieters, R., Lanz, M., Latokartano, J., Kämäräinen, J.K., 2020. AR-based interaction for human–robot collaborative manufacturing. Robotics and Computer-Integrated Manufacturing 63, 101891.

International Organization for Standardization, 2010a. Safety of machinery. General principles for design. Risk assessment and risk reduction. ISO Standard No. 12100:2010. https://www.iso.org/standard/51528.html.

International Organization for Standardization, 2010b. Basic human body measurements for technological design. Part 2: Statistical summaries of body measurements from national populations. ISO Standard No. 7250-2:2010. https://www.iso.org/standard/41249.html.

International Organization for Standardization, 2011. Robots and robotic devices. Safety requirements for industrial robots. Part 2: Robot systems and integration. ISO Standard No. 10218-2:2011. https://www.iso.org/standard/41571.html.

International Organization for Standardization, 2012. Safety of machinery. Risk assessment. Part 2: Practical guidance and examples of methods. ISO Standard No. 14121-2:2012. https://www.iso.org/standard/57180.html.

International Organization for Standardization, 2016. Robots and robotic devices. Collaborative robots. ISO Standard No. 15066:2016. https://www.iso.org/standard/62996.html.

Johannsmeier, L., Haddadin, S., 2016. A hierarchical human–robot interaction-planning framework for task allocation in collaborative industrial assembly processes. IEEE Robotics and Automation Letters 2, 41–48.

Kahanowich, N.D., Sintov, A., 2021. Robust classification of grasped objects in intuitive human–robot collaboration using a wearable force-myography device. IEEE Robotics and Automation Letters 6, 1192–1199.

Kana, S., Tee, K.P., Campolo, D., 2021. Human–robot co-manipulation during surface tooling: A general framework based on impedance control, haptic rendering and discrete geometry. Robotics and Computer-Integrated Manufacturing 67, 102033.

Kawasaki, A., Seki, A., 2021. Multimodal trajectory predictions for autonomous driving without a detailed prior map. In: Proceedings of the IEEE/CVF Winter Conference on Applications of Computer Vision, pp. 3723–3732.

Leng, J., Sha, W., Wang, B., Zheng, P., Zhuang, C., Liu, Q., Wuest, T., Mourtzis, D., Wang, L., 2022. Industry 5.0: Prospect and retrospect. Journal of Manufacturing Systems 65, 279–295.

Li, S., Zheng, P., Fan, J., Wang, L., 2021. Toward proactive human–robot collaborative assembly: A multimodal transfer-learning-enabled action prediction approach. IEEE Transactions on Industrial Electronics 69, 8579–8588.

Li, S., Zheng, P., Liu, S., Wang, Z., Wang, X.V., Wang, L., 2023. Proactive human–robot collaboration: Mutual-cognitive, predictable, and self-organising perspectives. Robotics and Computer-Integrated Manufacturing 81, 102510.

Liau, Y.Y., Ryu, K., 2022. Genetic algorithm-based task allocation in multiple modes of human–robot collaboration systems with two cobots. The International Journal of Advanced Manufacturing Technology 119, 7291–7309.

Liu, S., Wang, L., Wang, X.V., 2020a. Symbiotic human–robot collaboration: Multimodal control using function blocks. Procedia CIRP 93, 1188–1193.

Liu, X., Zheng, L., Shuai, J., Zhang, R., Li, Y., 2020b. Data-driven and AR assisted intelligent collaborative assembly system for large-scale complex products. Procedia CIRP 93, 1049–1054.

Liu, Y., Habibnezhad, M., Jebelli, H., 2022. Worker-aware task planning for construction robots: A physiologically based communication channel interface. In: Automation and Robotics in the Architecture, Engineering, and Construction Industry. Springer, pp. 181–200.

Magrini, E., Ferraguti, F., Ronga, A.J., Pini, F., De Luca, A., Leali, F., 2020. Human–robot coexistence and interaction in open industrial cells. Robotics and Computer-Integrated Manufacturing 61, 101846.

Mateus, J.C., Claeys, D., Limère, V., Cottyn, J., Aghezzaf, E.H., 2020. Base part centered assembly task precedence generation. The International Journal of Advanced Manufacturing Technology 107, 607–616.

Mauri, A., Lettori, J., Fusi, G., Fausti, D., Mor, M., Braghin, F., Legnani, G., Roveda, L., 2019. Mechanical and control design of an industrial exoskeleton for advanced human empowering in heavy parts manipulation tasks. Robotics 8, 65.

Mohammed, A., Schmidt, B., Wang, L., 2017. Energy-efficient robot configuration for assembly. Journal of Manufacturing Science and Engineering 139.

Mokhtarzadeh, M., Tavakkoli-Moghaddam, R., Vahedi-Nouri, B., Farsi, A., 2020. Scheduling of human–robot collaboration in assembly of printed circuit boards: A constraint programming approach. International Journal of Computer Integrated Manufacturing 33, 460–473.

Moon, J., Lee, B., 2018. Scene understanding using natural language description based on 3D semantic graph map. Intelligent Service Robotics 11, 347–354.

Ong, S., Yew, A., Thanigaivel, N., Nee, A., 2020. Augmented reality-assisted robot programming system for industrial applications. Robotics and Computer-Integrated Manufacturing 61, 101820.

Pang, G., Yang, G., Heng, W., Ye, Z., Huang, X., Yang, H.Y., Pang, Z., 2020. Coboskin: Soft robot skin with variable stiffness for safer human–robot collaboration. IEEE Transactions on Industrial Electronics 68, 3303–3314.

Patnayak, C., McClure, J.E., Williams, R.K., 2021. WASP: A wearable super-computing platform for distributed intelligence in multi-agent systems. In: 2021 IEEE High Performance Extreme Computing Conference (HPEC). IEEE, pp. 1–7.

Pedersen, M.R., Nalpantidis, L., Andersen, R.S., Schou, C., Bøgh, S., Krüger, V., Madsen, O., 2016. Robot skills for manufacturing: From concept to industrial deployment. Robotics and Computer-Integrated Manufacturing 37, 282–291.

Peternel, L., Fang, C., Tsagarakis, N., Ajoudani, A., 2019. A selective muscle fatigue management approach to ergonomic human–robot co-manipulation. Robotics and Computer-Integrated Manufacturing 58, 69–79.

Pulikottil, T.B., Pellegrinelli, S., Pedrocchi, N., 2021. A software tool for human–robot shared-workspace collaboration with task precedence constraints. Robotics and Computer-Integrated Manufacturing 67, 102051.

Realyvásquez-Vargas, A., Arredondo-Soto, K.C., García-Alcaraz, J.L., Márquez-Lobato, B.Y., Cruz-García, J., 2019. Introduction and configuration of a collaborative robot in an assembly task as a means to decrease occupational risks and increase efficiency in a manufacturing company. Robotics and Computer-Integrated Manufacturing 57, 315–328.

Rey, R., Cobano, J.A., Corzetto, M., Merino, L., Alvito, P., Caballero, F., 2021. A novel robot co-worker system for paint factories without the need of existing robotic infrastructure. Robotics and Computer-Integrated Manufacturing 70, 102122.

Romero, D., Bernus, P., Noran, O., Stahre, J., Fast-Berglund, Å., 2016. The Operator 4.0: Human cyber-physical systems & adaptive automation towards human–automation symbiosis work systems. In: IFIP International Conference on Advances in Production Management Systems. Springer, pp. 677–686.

Rosenberger, P., Cosgun, A., Newbury, R., Kwan, J., Ortenzi, V., Corke, P., Grafinger, M., 2020. Object-independent human-to-robot handovers using real time robotic vision. IEEE Robotics and Automation Letters 6, 17–23.

Sauer, V., Sauer, A., Mertens, A., 2021. Zoomorphic gestures for communicating cobot states. IEEE Robotics and Automation Letters 6, 2179–2185.

Schmidt, B., Wang, L., 2014. Depth camera based collision avoidance via active robot control. Journal of Manufacturing Systems 33, 711–718.

Shang, L., Wang, B., Yang, Z., Zheng, T., Destech Publicat, I., 2018. Research and implementation of the disassembling system of ship power equipment based on augmented reality. In: 2nd International Conference on Energy and Power Engineering, pp. 106–113.

Sheridan, T., 1986. Human supervisory control of robot systems. In: Proceedings. 1986 IEEE International Conference on Robotics and Automation. IEEE, pp. 808–812.

Tang, G., Webb, P., Thrower, J., 2019a. The development and evaluation of robot light skin: A novel robot signalling system to improve communication in industrial human–robot collaboration. Robotics and Computer-Integrated Manufacturing 56, 85–94.

Tang, K., Zhang, H., Wu, B., Luo, W., Liu, W., 2019b. Learning to compose dynamic tree structures for visual contexts. In: Proceedings of the IEEE/CVF Conference on Computer Vision and Pattern Recognition, pp. 6619–6628.

Tokadlı, G., Dorneich, M.C., 2019. Interaction paradigms: From human–human teaming to human–autonomy teaming. In: 2019 IEEE/AIAA 38th Digital Avionics Systems Conference (DASC). IEEE, pp. 1–8.

Unhelkar, V.V., Lasota, P.A., Tyroller, Q., Buhai, R.D., Marceau, L., Deml, B., Shah, J.A., 2018. Human-aware robotic assistant for collaborative assembly: Integrating human motion prediction with planning in time. IEEE Robotics and Automation Letters 3, 2394–2401.

Wang, H., Wang, W., Liang, W., Xiong, C., Shen, J., 2021. Structured scene memory for vision-language navigation. arXiv:2103.03454.

Wang, L., Liu, S., Liu, H., Wang, X.V., 2020a. Overview of human–robot collaboration in manufacturing. In: Proceedings of 5th International Conference on the Industry 4.0 Model for Advanced Manufacturing. Springer, pp. 15–58.

Wang, P., Zhang, S., Bai, X., Billinghurst, M., He, W., Sun, M., Chen, Y., Lv, H., Ji, H., 2019. 2.5DHANDS: A gesture-based MR remote collaborative platform. The International Journal of Advanced Manufacturing Technology 102, 1339–1353.

Wang, X.V., Wang, L., 2021. A literature survey of the robotic technologies during the COVID-19 pandemic. Journal of Manufacturing Systems.

Wang, X.V., Wang, L., Lei, M., Zhao, Y., 2020b. Closed-loop augmented reality towards accurate human–robot collaboration. CIRP Annals 69, 425–428.

Wang, Y., Sheng, Y., Wang, J., Zhang, W., 2017. Optimal collision-free robot trajectory generation based on time series prediction of human motion. IEEE Robotics and Automation Letters 3, 226–233.

Xu, W., Cui, J., Li, L., Yao, B., Tian, S., Zhou, Z., 2021a. Digital twin-based industrial cloud robotics: Framework, control approach and implementation. Journal of Manufacturing Systems 58, 196–209.

Xu, X., Lu, Y., Vogel-Heuser, B., Wang, L., 2021b. Industry 4.0 and Industry 5.0—inception, conception and perception. Journal of Manufacturing Systems 61, 530–535.

Yu, T., Huang, J., Chang, Q., 2021. Optimizing task scheduling in human–robot collaboration with deep multi-agent reinforcement learning. Journal of Manufacturing Systems 60, 487–499.

Yu, X., He, W., Li, Y., Xue, C., Li, J., Zou, J., Yang, C., 2019. Bayesian estimation of human impedance and motion intention for human–robot collaboration. IEEE Transactions on Cybernetics.

Zhang, H.B., Zhang, Y.X., Zhong, B., Lei, Q., Yang, L., Du, J.X., Chen, D.S., 2019. A comprehensive survey of vision-based human action recognition methods. Sensors 19, 1005.

Zhang, L., Guo, C., Zhang, L., Sheng, Y., 2021. A three-way human–robot task assignment method under intuitionistic fuzzy environment. In: 2021 IEEE 24th International Conference on Computer Supported Cooperative Work in Design (CSCWD). IEEE, pp. 587–592.

Zhang, R., Liu, X., Shuai, J., Zheng, L., 2020. Collaborative robot and mixed reality assisted microgravity assembly for large space mechanism. Procedia Manufacturing 51, 38–45.

Zhou, T., Wachs, J.P., 2018. Early prediction for physical human robot collaboration in the operating room. Autonomous Robots 42, 977–995.

CHAPTER 4

Mutual-cognitive and empathic co-working

In the present HRC, humans and robots operate collaboratively within shared workspaces, focus on ensuring workspace safety through continuous monitoring. However, these existing HRC systems is facing challenges related to the acquisition of high-level task knowledge for bi-directional intention understanding. To surmount these hurdles, this chapter introduces an approach based on MR and visual reasoning. This approach enables mutual-cognitive capabilities, thereby providing cognitive support to both human and robotic agents engaged in various tasks. The mutual-cognitive intelligence lies in several components, including dynamic environment perception, cognitive decision-making processes, and ergonomic robot control. These elements elevate the collaborative capabilities of human–robot teams. To illustrate the practical implications of this approach, the chapter presents a detailed demonstration involving the disassembly of an aging Electric Vehicle Battery (EVB). This case study underscores how human and robot operations can be combined to meet the collective needs of the team, thereby enhancing the overall efficiency of task execution.

4.1 Connotation

HRC in industrial settings aims for the collaborative efforts of humans and robots sharing a workspace (Wang et al., 2022). In this collaborative environment, task-related context awareness is essential for effective teamwork (Adeli et al., 2020). This includes ensuring human safety, facilitating active robot control, and enabling multimodal communication. Technologies such as CPS, DT, AR, and AI are emerging as solutions to perceive human–robot relationships in manufacturing activities. However, the current state of context awareness in HRC systems primarily operates at a non-semantic perception level, falling short of providing the mutual-cognitive intelligence and proactive collaborative knowledge required for optimal human and robot operations.

Mutual-cognitive intelligence in Proactive HRC involves empathic understanding between human–robot teams, encompassing task cognition, enhanced human cognition, and robot cognition. Task cognition involves generating co-working strategies through reasoned interpretation of human–robot–object relationships within the current workspace. This includes understanding current operation sequences, learning semantic knowledge regarding human intentions, and planning the next steps for the robot operation. Enhanced human cognition facilitates proactive communication within HRC systems, providing essential suggestions and support to humans for flexible decision-making. Robot cognition empowers robots to interact with humans based on ergonomic principles, optimizing factors such as handover positions and orientations to enhance human well-being.

In practice, the implementation of these concepts still face several algorithmic challenges, as depicted in Fig. 4.1. Task cognition entails two key challenges: 1) generating interpretable explanations for various scenarios and 2) devising task fulfillment strategies that guide both human and robot actions. To address these challenges, a dynamic Scene Graph (SG) construction approach can establish connections between on-site components and HRC tasks, enhancing contextual understanding of perceived objects. Subsequently, a Graph Convolutional Network (GCN)-based graph embedding method holds promise for generating cognitive task plans within HRC by leveraging task-specific context from the SG.

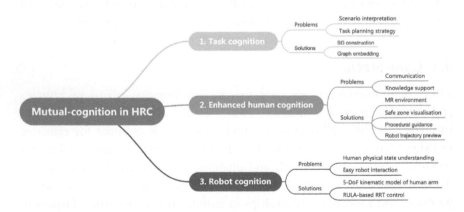

Figure 4.1 The connotation of mutual-cognition in Proactive HRC.

Enhancing human cognition in HRC entails addressing concerns such as: 1) establishing bi-directional communication between humans and robots and 2) providing domain knowledge support to humans when dealing with evolving HRC situations. During HRC deployment, a Mixed

Reality (MR)-based execution loop can be developed to facilitate seamless information exchange. The MR environment empowers human operators to access previously hidden information and anticipate future events, offering supports like safety zone visualization, procedural guidance, and previews of robot trajectories.

Requirements related to robot cognition include: 1) comprehending the physical state of humans through sequential actions and 2) planning optimal robot manipulation endpoints for convenient and comfortable interactions. Addressing these challenges may involve employing a 5-Degree-of-Freedom (5-DoF) kinematic model of the human arm alongside Rapid Upper Limb Assessment (RULA) rules to analyze human fatigue levels during collaboration. These analyzed parameters can then let Rapidly-exploring Random Tree (RRT) control algorithms adapt the position and orientation of the robot for optimal interaction.

In conclusion, to pave the way for mutual-cognitive HRC, this chapter introduces an MR-enabled, visual reasoning-based method. This mutual-cognitive task planning is derived from real-time SG knowledge on human–robot operational sequences. This knowledge is then integrated into the MR execution loop, allowing the robot to plan for human-desired manipulations while providing the human operator with intuitive guidance for manual operations. Simultaneously, the robot optimizes its interactive positions to align with ergonomic human postures, facilitating empathic teamwork.

4.2 A mixed-reality and visual reasoning-based framework

The system architecture for mutual-cognitive intelligence in HRC is depicted in Fig. 4.2, comprising three essential components: empathic collaboration in physical spaces, visual reasoning modules, and virtual replicas in cyber spaces, all underpinned by cognitive services in MR spaces. The fusion of physical and cyber spaces forms the HRC DT, responsible for updating physical system changes, offering previews of digital states, and facilitating collaborative decision-making. The HRC DT seamlessly integrates into the MR space to enable virtual-physical tracking and registration. In parallel, the MR system translates collaborative decisions, responsive to human–robot mutual operation needs and task properties, into cognitive services. These services enhance human flexibility, providing intuitive suggestions, and ensure proactive robot manipulation, including robot trajectory previews. This architectural approach enables mutual-cognitive, ergonomic, and empathetic HRC, prioritizing human well-being and sustaining production excellence in manufacturing tasks.

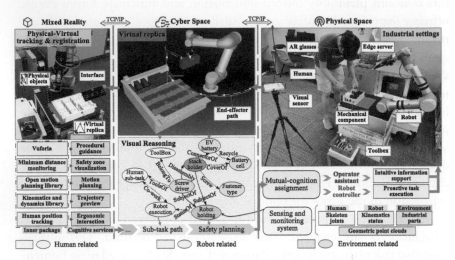

Figure 4.2 The overall framework of MR-enabled and visual reasoning-based mutual-cognition HRC.

Within the physical space, a sensing and monitoring system is developed to perceive human–robot states and detect changes in the surrounding environment. This involves the detection of human skeleton joints, industrial parts, and geometric point clouds, accomplished through Spatial Attention Pyramid Network (SAPNet) (Li et al., 2020), OpenPose (Cao et al., 2019), and OctoMap (Duberg and Jensfelt, 2020), respectively. These systems rely on data from visual sensors. An edge server, leveraging the ROS, collects data on robot status and facilitates feedback control on-site.

The cyber space component is responsible for translating physical HRC settings into virtual replicas, facilitating visualization and previews within the MR environment. Dynamic changes in human actions, robot operations, and task stages are communicated to digital HRC models. In return, robot path planning can be validated within these digital models before being executed in the physical world. A visual reasoning module is introduced in establishing relationships between humans, robots, the environment, and task structures. It derives co-working decisions from mutual-cognitive insights into HRC relationships throughout task processes. These decisions cater to the bi-directional needs of human–robot operations and dynamically allocate roles to humans and robots in HRC tasks.

Leveraging physical–virtual tracking and registration, MR-based cognitive services are made available within HRC systems. These services contain intuitive information support for human operators and proactive task execution for robots. For human operators, procedural guidance, delivered through various formats such as text, videos, and visualized operation sequences, is based on the Vuforia Toolbox. Additionally, the system visualizes a safety zone for human operation by continuously calculating the minimum distance between human and robot ontology. In the realm of robot control, the open motion planning library within ROS proactively devises robot motions for various task executions. Simultaneously, the kinematics and dynamics library achieves robot trajectory previews through a physical–virtual fusion process within the MR space. The system further analyzes the ergonomic risks associated with human skeleton poses through real-time human position tracking in OpenPose. It then plans ergonomic robot operations to facilitate effortless interaction. These cognitive services are dynamically delivered to human and robotic agents based on co-working decisions, achieving an empathetic teamwork environment between humans and robots.

4.2.1 Visual reasoning for mutual-cognition generation

The visual reasoning approach is utilized for understanding of the relationships between perceived objects. This approach facilitates HRC scene parsing from a mere perception level to a more advanced cognitive level. In previous work, Tang et al. (2019) introduced dynamic tree structures that were adept at capturing task-specific contexts essential for visual relationship and question answering. The work represented knowledge understanding visual relationships. Similarly, Kim and Lee (2019) proposed a model that used dynamic attention for focus transition. This model, grounded in human cognitive tendencies, exhibited a penchant for shorter reasoning paths. It generated attention maps that were more interpretable, contributing to the clarity of the reasoning process. Furthermore, the incorporation of scene graphs was proposed to gain structured knowledge of relationships between objects and their interactions (Shi et al., 2019). The introduction of scene graphs allows the ability to comprehend complex scenarios. In another approach, Venkatesh et al. (2020) leveraged both visual and language cues to enable robots to reason about object coordinates in picking and placing tasks. However, while these visual reasoning methods advanced the semantic understanding of diverse scenarios, these prior studies often overlooked the human–centric needs and the essential aspect of mutual-cognitive co-working support.

Driven by the motivation to cultivate empathic understanding within human–robot teamwork, a dynamic SG-based visual reasoning module takes center stage. This module assumes a pivotal role in deducing the operational requirements of both humans and robots as they collaborate in task completion. Ultimately, it allows the generation of mutual-cognitive co-working strategies for effective collaboration. As delineated in Fig. 4.3, this visual reasoning module encompasses four integral components:

a) (Scenario Perception) The phase involves object detection and the estimation of human body skeletons. These processes can locate industrial parts within the workspace and continuously track the movements of human skeleton joints.

b) (Temporal Node Updating) Leveraging the results of perceptual tasks, nodes associated with objects currently undergoing work-in-progress are activated, and their attributes are updated to reflect real-time changes in the environment.

c) (Dynamic Graph Construction) The subsequent step involves the dynamic construction of an SG. This graph connects the perceived objects (nodes) with the corresponding relations (edges), facilitating a comprehensive representation of the relationships within the workspace.

d) (Cognitive Strategy Mapping) SGs in different stages serve as the foundation for mutual-cognitive co-working strategies. These strategies are derived through the mapping of learned graph embeddings, which offer an interpretation of the ongoing human–robot operations.

This workflow illustrates the sequential procedures that underpin the visual reasoning approach, enabling mutual-cognitive intelligence and empathetic collaboration between humans and robots in dynamic industrial settings.

4.2.1.1 SAPNet-enabled object detection

An SAPNet is introduced to determine the coordinates and types of objects within HRC scenarios. The objects present in these scenarios encompass robot arms and work-in-progress workpieces. The SAPNet architecture, depicted in Fig. 4.4, is composed of four primary components: a feature extractor stem, a feature pyramid net, an output layer, and a loss function.

ResNet18, introduced in the work by He et al. (2016), serves as the foundational element for the feature extractor branch in our architecture.

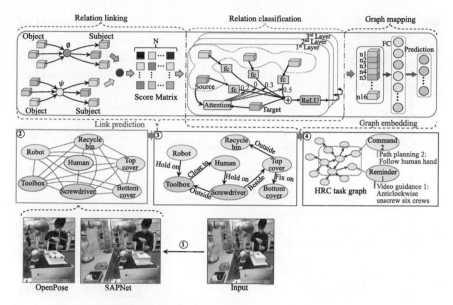

Figure 4.3 The workflow of visual reasoning-based mutual-cognitive strategy generation. (Adapted from Li et al. (2022).)

In the feature pyramid branch, a gradual integration of features from both high-level and low-level layers is performed, as outlined below:

$$b_i = \begin{cases} \mathrm{conv}_{(1 \times 1)}(f_i), & \text{if } i \leq 2, \\ B\_C(f_i), & \text{otherwise,} \end{cases} \tag{4.1}$$

$$g_i = \begin{cases} \mathrm{add}(b_i, \mathrm{upsamble}(b_{i+1})), & \text{if } i = 3, \\ \mathrm{add}(b_i, \mathrm{upsamble}(g_{i+1})), & \text{otherwise.} \end{cases} \tag{4.2}$$

Within the Feature Pyramid branch, we utilize a Flatten Bi-directional Long Short Term Memory (BiLSTM) layer, consisting of BiLSTM components whose detailed structure can be observed in Fig. 4.5. This layer processes feature maps at the pixel level, employing a sliding window of dimensions 3×3 pixels. This window collects information from the central pixel and its eight surrounding pixels, and subsequently outputs these nine values to augment the final channel of the feature maps. This process effectively enriches the information within the feature map. Importantly, this augmentation expands the feature map's dimensions from (B, H, W, C) to $(B, H, W, 3 \times 3 \times C)$, all while introducing no new training parameters into

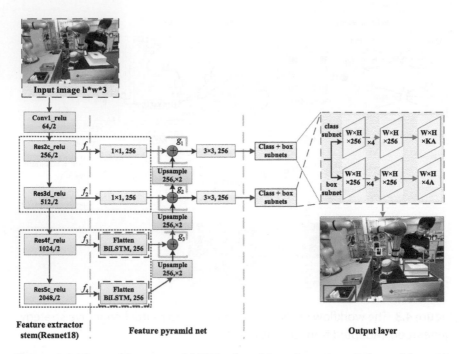

Figure 4.4 The architecture of SAPNet for object detection. (Adapted from Li et al. (2020).)

the detection model. Consequently, it does not exacerbate the complexity of the training process.

To facilitate the compatibility of the extracted feature maps with the input requirements of the BiLSTM, the reshaping of these feature maps is carried out. They are transformed into the format of $(B \times H, W, 3 \times 3 \times C)$, representing the data as a sequence. Subsequently, this sequential data is fed into the BiLSTM module, which, in turn, generates feature maps of dimensions $(B \times H, W, C)$ containing crucial information regarding the positional correlations of industrial parts.

Finally, the output feature map is restored to its original dimensions of (B, H, W, C) to ensure compatibility with the subsequent layers of the network. Here, we define C as the number of channels present in the feature maps. This multi-step process ensures that the extracted features are optimally prepared for the deeper layers of the network, maximizing the network's ability to understand and exploit the intricacies of the input data.

The utilization of the Flatten BiLSTM layer is primarily reserved for the deeper layers, specifically Res5c_relu and Res4f_relu, both of which

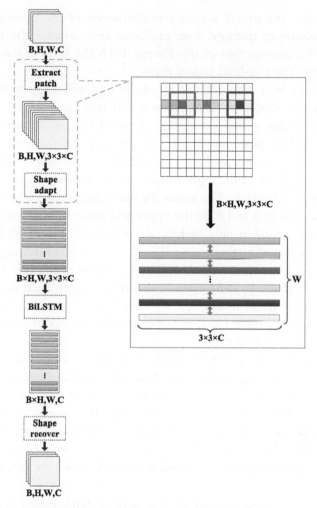

Figure 4.5 Spatial-attention layer with Flatten BiLSTM component. (Adapted from Li et al. (2020).)

are the output from ResNet18. In contrast, the application of this layer to shallow layers proves to be less effective in terms of capturing the intrinsic information present within images. This is due to the inherent limitations of shallow feature maps, which are characterized by larger sizes and a limited number of channels. Consequently, these shallow feature maps give rise to shorter data sequences for an increased number of basic units within the chained BiLSTM components. The challenges posed by these issues not only render the training process more arduous but also have the potential to compromise the overall accuracy performance of the networks. As a strategic measure to mitigate these problems and enhance the network's efficiency, the introduction of the Flatten BiLSTM layer is intentionally excluded from the low–level feature maps.

The output layer demonstrates simplicity and efficiency in its design, where it applies subnets for classification and regression to the final two upper feature maps within the feature pyramid branch. Each pixel, representing a sliding position, on the feature maps of these two subnets corresponds to a specific anchor. The anchors on the upper subnet correspond to 16^2 pixels of the input images, while the anchors on the lower subnet relate to 32^2 pixels. The values for the sliding steps of these anchors are configured to be 4 and 8 for the upper and lower subnets, respectively.

As a result, a total of nine anchors are generated at every sliding position. This is achieved by three scales ($[2^0, 2^{1/3}, 2^{2/3}]$) and three aspect ratios ($[2^0, 2^{1/3}, 1 + 2^{2/3}]$) for each anchor. These anchors encompass a range of scales, spanning from 16 to 82 pixels in the input images. Each anchor is assigned a one-hot vector of length K for classification targets and a 4-vector for box regression targets. Here, K represents the number of distinct classes associated with industrial parts.

In the classification subnet, the predicted probability of an object at each sliding position is determined by considering the nine anchors and K object classes. In the final stage of this process, sigmoid activations are introduced to produce a set of 9K binary predictions for each sliding location.

In the box regression subnet, the output consists of four components that predict the relative coordinates, represented as v, for the bounding box location of a given anchor. These relative coordinates v are determined as follows:

$$v_x = (c_x - c_x^a)/w^a, \quad v_y = (c_y - c_y^a)/h^a,$$
$$v_w = \log(w - w^a), \quad v_h = \log(h - h^a). \tag{4.3}$$

Here, we introduce $v = \{v_x, v_y, v_w, v_h\}$, which represents the set of relative predicted coordinates. In addition, we have $\{c_x^a, c_y^a, w^a, h^a\}$, denoting the center coordinates, width, and height of the anchor box. It is noting that these anchor box parameters can be pre-computed from the input image. Furthermore, we work with $\{c_x, c_y, w, h\}$, which are the predicted coordinates of the bounding box. To achieve the optimal parameters for the object detector, we employ a pair of loss functions, L_s^{cl} and L_v^{re}. These loss functions is used to quantify the errors in the subnets responsible for classification and box regression, respectively. The formulation of the loss is as follows:

$$L = L_s^{cl} + \lambda_{re} L_\nu^{re}. \tag{4.4}$$

The training process is influenced by two losses, and their relative importance is balanced by the parameter λ_{re} which is fixed at a value of 1. It is essential to acknowledge that in the image dataset, there exists a significant imbalance in the number of industrial parts belonging to different classes. Addressing this class imbalance issue during the training of the classification task, we employ the focal loss L_s^{cl}, expressed as follows:

$$L_s^{cl} = \begin{cases} -\alpha(1 - \hat{y})^\gamma \log \hat{y}, & \text{if } y = 1, \\ -(1 - \alpha)\hat{y}^\gamma \log(1 - \hat{y}), & \text{otherwise.} \end{cases} \tag{4.5}$$

The label of instances is represented by the variable y. To manage the importance of positive and negative examples, where positive examples are candidates generated by the network as potential objects, we use the parameter α. In our experimental setup, α is fixed at a value of 0.25. Additionally, we introduce the parameter γ to balance the weights assigned to easy and hard examples. This enables our training process to concentrate on challenging examples, which are instances that produce errors. In our experiments, γ is set to 2.

Traditionally, the loss function for regression tasks in object detection models relies on L_n-norms, which define the distance between the corners of two boxes. However, this approach encounters certain challenges. Firstly, the reliance on Intersection over Union (IoU) as an evaluation metric results in inconsistencies between training and testing. Secondly, the use of L_n-norms makes detection performance sensitive to the choice of IoU threshold, particularly when dealing with objects of varying sizes. To address these concerns, we implement the GIOU loss (Rezatofighi et al., 2019) in L_ν^{re}, which provides models with invariant accuracy regardless of the selected IoU threshold. The GIOU loss is calculated based on the corners of the predicted bounding box $\hat{R}$ represented as $\{x_1, y_1, x_2, y_2\}$ and the corners of the ground truth bounding box R^* denoted by $\{x_1^*, y_1^*, x_2^*, y_2^*\}$. The L_ν^{re} is calculated by

$$L_\nu^{re} = 1 - \left(\frac{|\hat{R} \cap R^*|}{|\hat{R} \cup R^*|} - \frac{(R^C - |R \cup R^*|)}{R^C} \right). \tag{4.6}$$

We can determine the area of the smallest enclosing box, denoted as R^C, by considering the coordinates of the bounding box B^C. These coordinates are given by

$$x_i^c = \min(x_i, x_i^*), \quad y_i^c = \min(y_i, y_i^*), \quad i = 1, 2,$$
$$R^c = (x_2^c - x_1^c) \times (y_2^c - y_1^c). \tag{4.7}$$

This architecture facilitates the identification and classification of objects in HRC scenarios, enhancing the system's ability to perceive and interact effectively with its environment.

4.2.1.2 Temporal node updating

The temporal node updating process involves identifying and tracking nodes that appear in HRC scenarios over time. These nodes correspond to detected objects, and their attributes are generated by the SAPNet, representing by position of a bounding box $v_i = [x_i, y_i, w_i, h_i]$ and a label $c_i \in \{1, \dots, k\}$. Here, k represents the number of object categories. Simultaneously, the human body skeleton is monitored in the images through OpenPose, and a similar approach is applied to define the location v and classes c of human hands. These temporal perceptual results are subsequently integrated into the following procedure and become active nodes denoted as V in the SG. The attributes of these nodes are continually updated using two sets of matrices, referred to as matrices $v \in \mathbb{R}^{n \times 4}$ and $c \in \mathbb{R}^{n \times k}$, as various objects are detected over time.

4.2.1.3 Link prediction for dynamic SG construction

To establish connections between perceived objects and their most relevant relations, i.e., node pairs, we introduce a link prediction approach. Each relation corresponds to an edge connecting a subject and an object within a node pair. The SG is then dynamically formed by establishing edges between these nodes. To optimize this process, we employ a two-layer perceptron, depicted in the lower-left corner of Fig. 4.3, to eliminate unnecessary node pairs. The degree of relatedness r_{ij} among $n \times (n-1)$ node pairs $\{x_i, x_j | i \neq j\}$ is defined as follows:

$$r_{ij} = f(x_i, x_j) = \langle \phi(x_i), \psi(x_j) \rangle, \quad i \neq j. \tag{4.8}$$

The relatedness function $f(\cdot, \cdot)$ is calculated through a matrix multiplication of $\phi(\cdot)$ and $\psi(\cdot)$. To achieve this, we employ a two-layer perceptron for the transformation process of x, generating $\phi(\cdot)$ and $\psi(\cdot)$ as outputs, respectively. The vector x encapsulates both object classes c and their spatial coordinates v. Subsequently, we apply a sigmoid function to r_{ij}, resulting in a relatedness score ranging from 0 to 1. We then identify the top K node pairs by ranking these relatedness scores in descending order. To refine the selection, we filter out node pairs that have more than half of their spatial regions overlapping with other nodes. The next step involves connecting the remaining node pairs with the appropriate relation types within the SG.

For the purpose of extracting contextual information between node pairs and predicting edge types within the SG, we introduce a three-layer attentional GCN. This process is illustrated in the lower-middle section (relation classification) of Fig. 4.3. Firstly, a linear transformation w is employed to derive features from neighboring nodes x_j related to a target node x_i. These features are then adjusted using weights α and combined, followed by activation through a nonlinear function σ, specifically the Rectified Linear Unit (ReLU). The propagation of these feature representations across GCN layers is represented as follows:

$$x_i^{(l+1)} = \sigma \left(x_i^{(l)} + \sum_j \alpha_{ij} w x_j^{(l)} \right). \tag{4.9}$$

To capture information from key node pairs, such as the robot node and a grasped object, the adjustment of attention to node features is essential and achieved by weights α. This allows focusing on specific node pairs that are most relevant. The attention mechanism between a target node x_i and its neighboring source node x_j can be computed using the following formula:

$$\begin{aligned} u_{ij} &= w_h^T \sigma(W_a[x_i, x_j]), \\ \alpha_{ij} &= \text{Softmax}(u_{ij}). \end{aligned} \tag{4.10}$$

Here, w_h and W_a represent the parameters of a two-layer perceptron. Utilizing the acquired node pairs and their corresponding relation types, we proceed to dynamically construct an SG by establishing edges E between nodes V, as depicted in the upper-right corner of Fig. 4.3.

4.2.1.4 Graph embedding for cognitive strategy mapping

Having constructed an SG based on the perceived objects, the subsequent stage involves learning graph embeddings and mapping them into human reminders and robot commands, thus forming mutual-cognitive task strategies. As previously mentioned, node pairs within the SG encapsulate implicit interpretations of human–robot teamwork. For instance, a pair consisting of a human node and a manipulated industrial part may convey human operational intentions, while a pair involving a human node and a robot node might signify potential contact hazards.

In this regard, skip-connect edges are introduced among all nodes to directly exchange information between nodes. The SG encompasses three distinct types of connections: from *subject* to *relation*, from *relation* to *object*, and from *object* to *object*. To effectively extract feature representations across these diverse connections, a three-layer attentional GCN is employed, as depicted in the relation classification section of Fig. 4.3.

Utilizing the matrix $X \in \mathbb{R}^{d \times T_n}$ to represent neighboring nodes x_j, Eq. (4.9) can be reformulated as shown in equation $x_i^{(l+1)} = \sigma(WX^{(l)}\alpha_i)$, where d denotes the dimension, and T_n represents the quantity of x_j. Employing this notation, the feature transformation among GCN layers for nodes is defined as follows:

$$X_i^o = \sigma(\overbrace{W^{skip}X^o\alpha^{skip}}^{\text{Other nodes}} + \overbrace{W^{sr}X^r\alpha^{(sr)} + W^{or}X^r\alpha^{(or)}}^{\text{Neighboring relations}}). \qquad (4.11)$$

Here, s corresponds to the *subject*, r pertains to the *relation*, and o represents the *object*. The first part in Eq. (4.11) depicts to the features of skip-connected nodes, while the latter part relates to neighboring relations. Similarly, the representations of relations are propagated as described below:

$$X_i^r = \sigma(X_i^r + \overbrace{W^{rs}X^o\alpha^{(rs)} + W^{ro}X^o\alpha^{(ro)}}^{\text{Neighboring nodes}}). \qquad (4.12)$$

The final step in this process is graph mapping, illustrated in the lower right corner of Fig. 4.3. Here, a fully connected layer is added on top of the three-layer attentional GCN, enabling a linear transformation of the extracted feature representations. Ultimately, a Softmax function is employed in conjunction with the fully connected layer to facilitate the learning of graph embeddings, mapping them to corresponding human reminders and robot commands.

The construction of the SG and the embedding process are acquired through three supervised training procedures:

a) For relation linking, a binary cross-entropy loss is employed during the training phase.

b) In the case of relation classification, a multiclass cross-entropy loss is utilized.

c) Concerning graph mapping, two additional multiclass entropy losses are formulated to guide the mapping of human reminders and robot commands, respectively.

4.2.1.5 An alternative zero-shot multi-feature fusion method

In the context of tasks that are similar yet distinct, many graph construction methods require the acquisition of new knowledge representations in order to generate suitable strategies for both human and robotic operations. This situation imposes limitations on the applicability of HRC in real-world scenarios. To bridge this gap in research, an alternative approach is explored – one that introduces a zero-shot multi-feature fusion method to allocate mutual-cognitive teamwork strategies for human and robotic agents as they progress through the task.

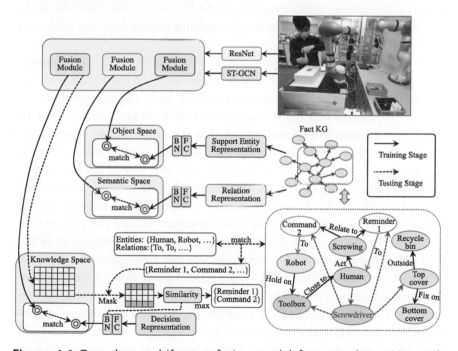

Figure 4.6 Zero-shot multifeature fusion model for mutual-cognitive task planning. (Adapted from Li et al. (2023).)

Illustrated in Fig. 4.6, the workflow of mutual-cognitive HRC systems start from visual input. This input undergoes processing via the Spatial-Temporal Graph Convolutional Network (ST-GCN) model (Yan et al., 2018) for human action recognition and the ResNet (Chen et al., 2021) algorithm for object detection. Subsequently, three fusion modules are deployed, each operating within a distinct space: object, semantic, and knowledge. In the object space, the objective is to establish mappings between perceived objects and recognized actions, effectively integrating them into a KG. The semantic space, on the other hand, focuses on the learning of relational representations within the KG. Finally, the knowledge space plays a pivotal role in driving decision-making, offering collaborative strategies for the HRC system. The decision-making process fuses information of both the object and semantic spaces within the KG. Furthermore, to facilitate the method's capacity to deduce rational decisions in scenarios that deviate from the norm – such as the absence of perceived objects or the presence of an entirely unfamiliar KG – a masking mechanism is introduced. This mechanism tackles the alignment process of the KG by maximizing the similarity of decision representations. In order to accommodate the multiple feature spaces and the introduced masking mechanism, the following aspects are brought into focus.

In the object space, our initial focus is on constructing connections. These connections can link the support entities within the Fact KG to the perceived outcomes. These outcomes include both the detected objects ($O = \{o_1, o_2, \ldots, o_n\}$) and the recognized actions ($A = \{a_1, a_2, \ldots, a_n\}$). This results in the formation of associations that directly establish connections between the observed results and the KG. Upon this foundation, we shift our attention to the semantic space. Here, our aim is to acquire a understanding of the relations between the extracted node features, represented as $\{(o, a)|o \in O, a \in A\}$. This leads to the establishment of relevant triplets within the KG, enriching its content and structure. Besides, it is imperative to fuse the role of recognized human actions (a) in the construction of the KG. The recognition of actions directly informs and influences potential teamwork strategies within HRC tasks. It serves as a direct representation of human operational intentions for collaborative decision-making.

Then, the knowledge space is used for the creation of connections between collaboration strategies ($S = \{s_1, s_2, \ldots, s_n\}$) and the perceived ($o, a$) pairs. Here, we employ a fusion feature extractor, denoted as $F_\theta(o, a)$, to facilitate the integration of multimodal information, bridging the gap between human actions (o) and the detected industrial parts (a). Simultaneously, we introduce $G_\phi(s)$ as the decision representation governing the task planning strategy, referred to as s.

To unify the features of strategies (s) with those of the pair (o, a), we enlist the Probabilistic Model of Compatibility (PMC). This mechanism can ensure that the features align cohesively and that they are well-suited for collaborative decision-making, as elucidated below:

$$P(s|o_n, a_n) = \frac{\exp\left(F_\theta(o_n, a_n)^T G_\theta(s)/\tau\right)}{\sum_{s' \in s} \exp\left(F_\theta(o_n, a_n)^T G_\theta(s')/\tau\right)}. \tag{4.13}$$

Here, the symbol $\sum$ encompasses all conceivable connections within the Fact KG. Alongside this, we introduce the parameter τ, which serves as a loss temperature to facilitate the optimization of our model. As we delve into the process of fine-tuning the learning parameters in the PMC, the formulation of the loss function unfolds as outlined below:

$$l_s = -\sum_n^N \sum_{s' \in s} \varphi(s, \hat{s}) \log P(\hat{s}|o_n, a_n). \tag{4.14}$$

The function $\varphi(s, \hat{s})$ is of binary indicator. It evaluates as "true" when the predicted strategy $\hat{s}$ in the training process matches the correct strategy s, and conversely, it evaluates as "false," represented by 0 when there is a mismatch.

The alignment of detected objects (o), recognized actions (a), and the KG is realized through these three feature spaces. This is followed by the application of Batch Normalization (BN) and a Fully Connected (FC) layer, as depicted in Fig. 4.6. The relationships (r) learned between support entities within the KG can be applied in the subsequent masking process. After undergoing three fusion modules, the embeddings of decisions within the KG are represented as $G_\phi^r(s)$ and $G_\phi^e(s)$ for relations and entities, respectively. Similarly, the embeddings for the pairs (o, a) are defined as $F_\theta^r(o, a)$ and $F_\theta^e(o, a)$, which are essential components for the masking process.

In real-world scenarios, several situations can introduce error decisions during HRC task planning. These situations include instances where there are missing perceived objects (such as the red nodes and arrows in Fig. 4.6), errors in the perception of objects, and the introduction of new, yet analogous scenarios. These unforeseen circumstances can impede the development of mutual-cognitive intelligence within HRC. To address these challenges, we introduce a zero-shot learning algorithm that leverages a masking mechanism. This mechanism empowers the approach to make accurate collaborative decisions even when faced with unobserved perceptual outcomes.

The embeddings $F_\theta^r(o, a)$ and $F_\theta^e(o, a)$, derived from the fusion of object and semantic feature spaces for the matching of entities and relations within the KG, are illustrated in Fig. 4.6. Given the perceived joint features (o, a), the similarity of relations (r) is determined by $F_\theta^r(o, a)^T G_\phi^r(s)$, whereas the similarity of entities (e) is computed using $F_\theta^e(o, a)^T G_\phi^e(s)$. Subsequently, we establish the candidate entity set (C_e) by selecting the top k_e similarity values among these entities (e). In a parallel process, the candidate relation set (C_r) is formed by retaining the top k_r similarity values among the relations (r). Concurrently, the potential decision set (C_s) is constituted through the following procedure:

$$c_s = \{\bar{s} | (\exists (\bar{s}, r, e) \vee \exists (e, r, \bar{s})) \wedge r \in c_r \wedge e \in c_e\}. \tag{4.15}$$

Subsequently, we introduce the mask mechanism to facilitate the computation of similarity values for all decisions within the set, denoted as $s_n \in S$. This mechanism operates as follows:

$$\text{sim} ((o, a), s_n) = \begin{cases} \frac{F_\theta(o,a)^T G_\phi(s_n)}{\tau}, & \text{if } s_n \in c_s, \\ \frac{F_\theta(o,a)^T G_\phi(s_n)}{\tau+\delta}, & \text{otherwise.} \end{cases} \tag{4.16}$$

Here, the symbol δ means the masking score. To conclude the process, we calculate the deduced decisions for task planning within HRC systems as expressed below:

$$\hat{s} = \arg \max_{s_n \in s} \text{sim} ((o, a), s_n). \tag{4.17}$$

4.2.2 Safe and ergonomic robot motion planning

Using task planning strategies derived from the visual reasoning module, a robot can carry out operations that are aligned with human requirements within a shared workspace. As illustrated in Fig. 4.7, to enhance human well-being and ensure the mutual-cognitive capabilities in their collaborative efforts, the robot follows a set of standards that encompass safety, ergonomics, and proactivity. These standards are realized through the integration of three key modules:

a) (Real-time Collision Avoidance) Firstly, a real-time collision space is constructed based on RGB-D data (combining RGB images with depth information) collected from the on-site workspaces. This collision space provides safety constraints for the generation of robot action trajectories.

b) (Ergonomic Interactive Actions) The system incorporates interactive actions that facilitate human–robot cooperation (e.g., handovers) while

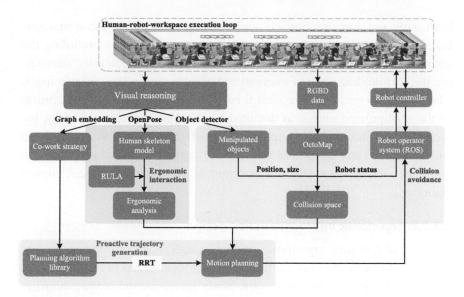

Figure 4.7 The workflow of robot motion planning.

adhering to ergonomic requirements designed to reduce worker fatigue.

c) (Proactive Trajectory Generation) Taking into account the aforementioned considerations and the assigned robot tasks, a rapid and robust motion planning algorithm is employed to proactively generate robot trajectories that align with the specified objectives.

The subsequent sections will provide detailed methodologies related to robot control.

4.2.2.1 Collision avoidance based on real-time obstacle space

Ensuring the safety of both human workers and robots necessitates the identification of regions where contact hazards may arise, thereby defining obstacle spaces that robots must avoid during their movements. These obstacle spaces encompass potential collisions with both static obstacles (e.g., tables) and dynamic obstacles (e.g., human bodies). The establishment of a real-time obstacle space in the motion planning process involves three-step procedures. Firstly, a 3D occupancy grid mapping technique known as OctoMap is employed to create a representation of the obstacle space within HRC systems. Leveraging RGB-D data from the on-site workspace, the OctoMap algorithm continuously updates a real-time 3D map that includes both static and dynamic obstacle spaces. Then, to facilitate robot manipulation of the target object, the obstacle space is adjusted by excluding the manipulated object from consideration. This step is executed by eliminating the surrounding areas of manipulated objects within the obstacle space. The process to remove these areas is based on the position and dimensions of the manipulated object, as detected by the object detector. Finally, information regarding the kinematics of robots is extracted from the ROS and is subsequently integrated into the 3D map. This integrated map can highlight collision-prone regions that must be avoided during robot motion planning.

4.2.2.2 Ergonomic interactive action design

The goal of designing ergonomic interactive actions is to enhance the comfort and safety of teamwork for human operators engaged in HRC systems. These interactive actions encompass direct interactions between human–robot agents and the handover of manipulated objects, both of which are common operations in HRC tasks. Achieving ergonomic interactions involves defining the robot's interactive space, which includes factors such as the position and orientation of a handover point. The design of this interactive space adheres to ergonomic criteria, i.e., RULA (McAtamney and Corlett, 1993), to ensure that it is easily and comfortably accessible by human hands. These criteria contain the following rules:

a) (Range of Upper Arm Movement) The interactive space allows for a range of movement of the upper arm, ranging from a 20° extension to a 20° flexion.

b) (Range of Lower Arm Movement) It accommodates the range of motion of the lower arm, which extends to a 60°–100° flexion.

c) (Wrist Position) The wrist is maintained in a neutral position to minimize strain.

In accordance with these requirements, a 5-Degree-of-Freedom (5-DoF) kinematic model of the human arm is employed to define the robot's interactive space, as depicted in Fig. 4.8. This model considers several key joints and parameters:

a) (Shoulder Joints) These include three degrees of freedom, specifically represented as shoulder adduction R_A, shoulder flexion R_F, and shoulder rotation R_R.

b) (Elbow Joint) Defined as a single degree of freedom joint as R_E.

c) (Wrist Joint) Referred as joint R_W.

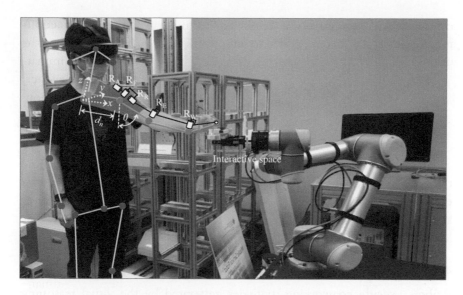

Figure 4.8 5-DoF kinematic model of a human arm in Proactive HRC.

The coordinates of these joints i are represented as p_i, and the corresponding joint angles are denoted as θ_i. Linkages represent the upper arm, lower arm, and hand. A three-dimensional Cartesian coordinate system is established, with the human neck point n serving as the origin. The coordinate system aligns with the body's relative orientation, where the Y-axis corresponds to the forward direction, the X-axis represents rightward, and the Z-axis points upward. Using forward kinematics of the linkage, the coordinates of the palm, representing the human–robot interactive space, can be derived through the following equation:

$$p_h = d_n + A_a A_f A_r (d_h + A_e(d_f + A_w d_w)). \qquad (4.18)$$

The calculation of the above equation involves several key parameters: d_n represents the distance from the neck to the shoulder; d_w means the distance from the wrist to the palm; d_h denotes the length of the upper arm; d_f represents the length of the lower arm; and A_i corresponds to the rotation matrix of joint i.

The values of d_n, d_h, d_f, and d_w are determined through the real-time estimation of human skeleton joints. The upper arm's rotation angle is represented as $\theta = \arccos(\cos\theta_a \cos\theta_f)$, considering that joints θ_a and θ_f are mutually perpendicular. The range of rotation for the lower arm is indicated as θ_e and θ_w signifies the wrist's range of motion. To adhere to RULA requirements, it is recommended to set specific rotation ranges for the joints of the human arm:

$$0° \leq \theta \leq 20°,$$
$$60° \leq \theta_e \leq 100°, \tag{4.19}$$
$$\theta_w \approx 180°.$$

Using the forward kinematic equation calculation, we can determine the human–robot interactive space. Subsequently, the robot positions its end-effector within this interactive space, enabling ergonomic handover actions that align with the requirements for comfortable human–robot collaboration.

4.2.2.3 Motion planning for proactive trajectory generation

Robots execute co-working strategies generated by the visual reasoning module through a proactive motion planning algorithm. These strategies involve tasks like object manipulation, pick-and-place operations, or handovers. Specifically, the RRT (LaValle, 1998) algorithm is employed for planning continuous robot trajectories from their initial positions to the desired endpoints.

To ensure collision avoidance, the RRT algorithm builds a tree structure, starting from the initial robot positions and extending toward ergonomic interactive points. It does so by generating random samples within the configuration space. With each new sample, the algorithm attempts to establish connections with the nearest existing tree points. Only connections that adhere to constraints are accepted, and the resulting points are integrated into the growing tree. Ultimately, this process yields a dynamic path from the starting positions to the ergonomic interactive destination, enabling the robot to perform its tasks without encountering collisions.

4.3 Case study

In this section, we implement a prototype system for mutual-cognitive HRC, focusing on the disassembly task of EVBs. We then assess the generation of cognitive task planning strategies using the visual reasoning module. Subsequently, we evaluate the mutual-cognitive intelligence in Proactive HRC from experimental outcomes related to intuitive human support and the robot's safe, ergonomic, and proactive manipulation.

4.3.1 Mutual-cognitive HRC for disassembly of EVBs

The disassembly of EVBs requires solutions in the context of the growing electric vehicle industry. On the one hand, certain flexible subtasks, such as cutting wires and removing glue, necessitate manual human operations. On the other hand, robot manipulation is indispensable, particularly for high-risk tasks like picking and placing battery cells. Hence, HRC offers an effective solution for EVB disassembly, leveraging the strengths of both humans and robots.

The system configuration for mutual-cognitive HRC is depicted in Fig. 4.9. This setup encompasses on-site equipment, an edge server, cloud server, ROS platform, and robot controller, all enclosed within an MR execution loop. Within this loop, the Azure Kinect camera is employed to capture real-time on-site images as the disassembly task progresses. The edge server handles the estimation of human skeleton joints, industrial parts, and cloud points within the workspace at various stages of the EVB disassembly. These perceptual results are dynamically assembled into an SG through the cloud server's visual reasoning module.

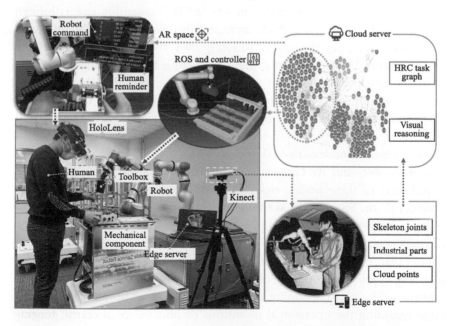

Figure 4.9 Prototype system setup of Proactive HRC for EVB disassembly task.

The SG is used as a dynamic connector, linking humans, robots, and their operational knowledge across different phases of the EVB disassembly task. This interconnected knowledge includes video guidance for human operations and robot path planning. The video guidance is presented on MR glasses (HoloLens 2), providing human operators with step-by-step instructions on EVB component removal. Simultaneously, path planning commands are transmitted to the ROS and a robot controller.

As a result, a mobile robot (comprising a Universal Robots UR5 and MiR100 base) can proactively engage in interactive actions with humans or undertake hazardous subtasks, such as the precise picking and placing of battery cells. With the timely support of reminders for human operators and robot commands, both human and robotic agents collaborate in a mutual-cognitive manner to successfully complete the EVB disassembly process.

4.3.2 Visual reasoning for co-working strategy generation

The visual reasoning module is used to generate task planning strategies throughout the various subtasks involved in EVB disassembly.

4.3.2.1 HRC SG dataset for EVB disassembly

Within our laboratory environment, the EVB disassembly task is broken down into 11 substages, which include the delivery of tools, unscrewing screws, opening the cover, handing over the cover, testing electric power, cutting wires, removing glues, recycling PCB modules, recycling Thermo sensors, recycling Ion cells, and organizing the toolbox. To provide a comprehensive illustration of the EVB disassembly, the complete graph representing HRC settings is displayed in Fig. 4.10. This dataset, containing knowledge related to the disassembly processes, comprises six distinct types of nodes: *Tool*, *Workpiece*, *Human*, *Reminder* for manual operation, *Robot*, and its *Command*. With respect to the labels and spatial coordinates of these nodes, eight different relations can be established to connect them. Specifically, as the disassembly task progresses, a total of 11 SGs can be generated, featuring various nodes and edges that are activated through temporal visual input. These 11 disassembly stages encompass a dataset that includes 11 forms of video guidance and 10 distinct robot path planning approaches, offering operational knowledge to address a range of scenarios. The visual reasoning approach is geared toward acquiring knowledge regarding the operational intentions of human–robot teams, forging relevant connections between human reminders and robot commands to facilitate their cognitive cooperation in the disassembly process.

Moreover, the dataset contains 779 RGB images with corresponding depth information. These images show the presence of a human opera-

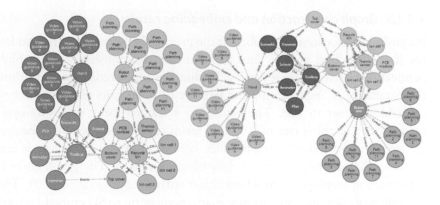

Figure 4.10 HRC SG in the EVB disassembly task.

tor, along with 13 diverse industrial parts: Toolbox, Screwdriver, Ammeter, Plier, Scissor, Hammer, Recycle bin, Top cover, Bottom cover, PCB module, Thermo sensor, Ion cell, and Robot arm. To provide annotations for this dataset, each industrial part is labeled with its assigned category and four coordinates defining a bounding box. Furthermore, the dataset also includes annotations that describe the relations between industrial parts in each image.

4.3.2.2 Scenario perception results

Accurate object detection in HRC scenarios serves as a critical prerequisite for the generation of nodes and the dynamic construction of SG to map graph embedding to mutual cognition. In our experimental setup, we have partitioned the dataset into two segments: a training set comprising 467 images and a testing set comprising 312 images. The ResNet 18 model is configured with a total of 13 categories denoted as k. For the training of the object detector within SAPNet, a Tesla V100 GPU (16G) is utilized, with the optimizer being Stochastic Gradient Descent (SGD).

Furthermore, OpenPose is harnessed to estimate the coordinates of 18 body skeleton joints from the images, thereby enabling the detection of human hands and 13 distinct industrial parts in the scenario perception phase. This initial detection process results in the generation of perceptual outcomes. Leveraging temporal visual input, diverse objects are accurately located and appropriately categorized, facilitating the dynamic generation of nodes within the SGs.

4.3.2.3 Graph construction and embedding results

Regarding the locations and labels of the previously perceived object nodes, the link prediction process aims to learn their relationships and establish connections between these nodes. Within the link prediction algorithm, the parameter K, representing the number of node pairs with the highest relatedness, is set to 128. The algorithm is trained using the SGD optimizer with a learning rate of 0.001. In terms of graph embedding, features are extracted from the 14 graph nodes, consisting of one human node and 13 industrial part nodes, using a fully connected layer. The graph embedding training employs the SGD optimizer with a learning rate of 0.01. The training processes, spanning from scenario perception to SG embedding, are executed on a Tesla V100 GPU. During testing, the trained model perceives various objects across different disassembly stages, dynamically connects relationships among these objects to form an SG, and triggers video guidance as human reminders and path planning as robot commands.

To illustrate the visual reasoning module, Fig. 4.11 showcases three examples of co-working strategy generation during the stages of unscrewing screws, testing electric power, and recycling PCB modules, respectively. In the upper side of Fig. 4.11, the mutual–cognitive decision is utilized to assist the HRC disassembly system while the screws are being unscrewed in a collaborative manner. The human operator is provided with video guidance on the procedure of anticlockwise unscrewing of six screws. Meanwhile, the robot is programmed to follow the movements of the human hand. Despite the absence of certain industrial parts in the SG that serve as supporting entities, the algorithm can infer appropriate teamwork strategies by identifying the decision with the highest similarity values. Consequently, the visual perception model, which includes ResNet and ST-GCN algorithm, does not require task relearning when confronted with a comparable yet distinct scenario. In the middle section of Fig. 4.11, an SG that indicates human–robot teamwork states during the testing electric power stage is created by connecting relationships among different perceived objects in this phase. The graph embedding for this scenario is translated into task planning strategies. The human operator receives intuitive video guidance for testing the electric power of three Ion cells, while the robot aligns with the human's hand motion, especially if the human changes tools. Subsequently, as the human grasps a pair of pliers in the next stage, the visual reasoning module infers the human–robot operational intentions of loosening a PCB module. The bottom section of Fig. 4.11 illustrates the delivery of video guidance to the human operator for removing glue from the PCB module.

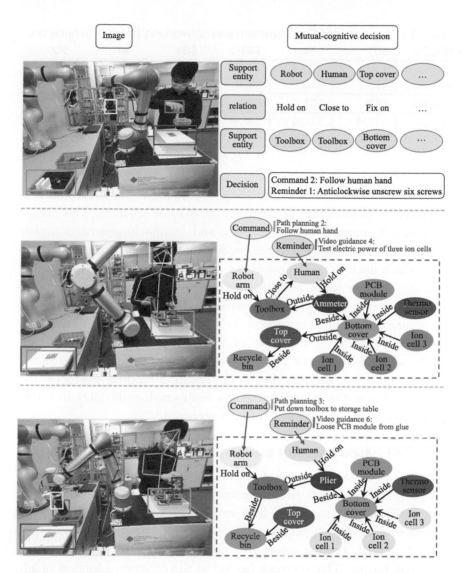

Figure 4.11 Samples of decision making in Proactive HRC by SG methods.

The robot places the toolbox on a storage table and proceeds with recycling the PCB module. This collaborative approach ensures that human and robotic agents understand each other's operational needs and proactively perform actions aligned with their partner's requirements at each stage of the EVB disassembly task.

Table 4.1 Accuracy of SG construction and co-working strategy generation.

HRC SG	SG1	SG2	SG3	SG4	SG5	SG6
SGGen+	13/14	14/14	21/21	20/21	34/34	32/34
Precision	91.67	95.00	96.36	91.67	94.44	95.74
HRC SG	**SG7**	**SG8**	**SG9**	**SG10**	**SG11**	
SGGen+	34/34	33/34	34/34	33/34	32/34	
Precision	96.30	92.86	96.00	93.75	92.31	

Beyond the demonstration, the SGGen+ (Yang et al., 2018) metric is employed to evaluate the performance of SG construction. This metric calculates the number of nodes, edges, and accurately generated triplets within the SGs. Additionally, the prediction precision of graph embedding, which maps to mutual cognition, is assessed across 11 SGs. Table 4.1 presents the results, showcasing that the link prediction approach effectively connects correct relations between nodes. In the second row of Table 4.1, the accuracy of graph construction X/Y is evaluated, where X represents the correctly predicted results out of Y total nodes and edges within an SG. Finally, the last row in Table 4.1 assesses the precision of co-working strategy generation through the visual reasoning module. Based on the constructed SGs, the mapping processes from graph embedding to HRC mutual–cognitive strategies achieve high accuracy.

4.3.3 MR-based operator assistance and robot control

MR in the context of manufacturing seamlessly merges DT models with AR environments. While AR concentrates on visually blending virtual objects with the physical world, MR extends beyond this by offering the analysis of a system's physical states, the simulation of future conditions using DT models, and the subsequent presentation of this simulation data in an AR format. This approach has led to the extensive adoption of advanced MR technologies in HRC (Wang, 2022). A notable example of this integration is showcased in the work by Hietanen et al. (2020). They developed an HRC system incorporating a projector and wearable MR glasses. This MR interface provided human operators with real-time insights into the robot's status and evolving safety zones within the shared workspace. For instance, an MR-based execution loop allowed human operators to seamlessly communicate the assembly status to the robot, without necessitating expertise in robotics. In this way, MR-based communication allows for seamless information exchange between the two participants, while also offering intuitive domain knowledge support for operator assistance.

During the EVB disassembly process, the co-working strategy generated is transmitted to human–robot teams via MR glasses, enhancing the human operator's capabilities, as depicted in Fig. 4.12. These MR glasses provide human operators with intuitive support, including procedural guidance with a fusion of virtual and physical elements, video guidance for manual operations, and real-time visualizations of safety zones. The MR glasses also offer a preview of the robot's trajectory before execution, enabling human operators to anticipate the robot's next intended motion. Therefore, human operators benefit from increased flexibility and cognitive support, aiding their decision-making regarding subsequent disassembly operations based on MR environment suggestions.

Conversely, the work of Bottani and Vignali (2019) used MR techniques to let humans directly guide and instruct robot manipulations. In this approach, users were granted the ability to define 3D points for robot paths through the use of an MR interface (Ong et al., 2020). Hernández et al. (2020) explored robotic motion planning to cater to users' high-level requests for robot manipulation, steering away from the need for explicit low-level directives. These MR-based robot programming methods are feasible for Proactive HRC systems. The system can dynamically plan proactive robot motions. Furthermore, within the MR environment, robot motions can be simulated as an HRC DT component of the collaborative strategy. Subsequently, motion planning commands can be transmitted to the robot, facilitating proactive task execution. This section evaluates robot task execution performance from three critical aspects: feasibility, safety, and ergonomics. Feasibility analysis focuses on validating the system's capacity to perform specified functions, such as determining if the robot can generate a path for a given task. Safety analysis assesses the robustness of obstacle detection and collision avoidance within the system. Ergonomic analysis aims to ensure that the interactive points determined by the system are comfortably reachable by human hands. These considerations are paramount when the robot collaboratively disassembles EVBs. To visualize the robot's motion planning process with a focus on these three concerns, a physical-simulation platform, Gazebo, is employed as the HRC DT.

Feasibility Test. In response to commands from the visual reasoning module, the HRC system generates corresponding trajectories for the robot to execute necessary human-assisted operations. For demonstration, a common subtask involving the robot grasping, moving, and delivering a

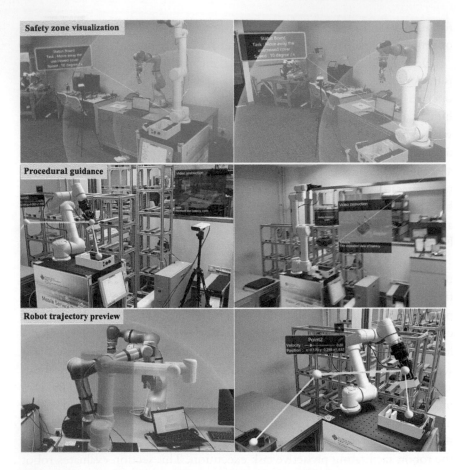

Figure 4.12 System demonstration in the MR environment.

toolbox to a position accessible to human partners is utilized. Figs. 4.13a and 4.13b showcase an execution stage of this subtask, while the entire trajectory generated is visualized in the same figure.

Safety Test. Collision avoidance is a fundamental requirement in HRC systems. Employing the same robot subtask, the planning module can generate a safe trajectory that avoids collisions, ensuring the safety of both human and robotic agents. Fig. 4.13c illustrates a scenario where an obstacle is detected and introduced into the workspace between two agents, obstructing the robot arm's movement. With the obstacle space dynamically updating based on perceptual results, the robot adeptly navigates around these obstacles to generate a safe trajectory, as depicted in Fig. 4.13d.

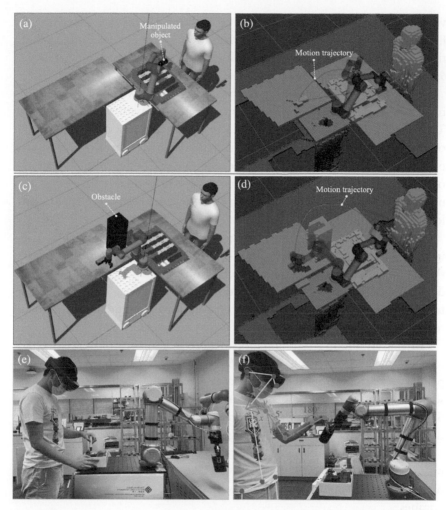

Figure 4.13 Test of robot motions under feasibility, safety, and ergonomic aspects.

Ergonomic Test. This test assesses the comfort and convenience of interactive actions between human–robot teams. When a human operator adopts various postures, the robot identifies suitable handover positions that the human can easily reach. As exemplified in Fig. 4.13e, the HRC system learns that the human operator is in the process of opening the top cover of EVBs and requires an Ammeter for testing the electric power of Ion cells. Consequently, the robot calculates an ergonomic handover position and delivers the Ammeter to the human based on assigned commands. Fig. 4.13f visually depicts the designated handover position for the Ammeter transfer from the robot to the worker. These handover points are determined using the forward kinematic equation, with specific parameters $\theta_a = 5°$, $\theta_f = 5°$, $\theta_r = 0°$, $\theta_e = 80°$, and $\theta_w = 180°$.

A total of 10 participants took part in the test, with each individual instructed to position themselves around the robot within the workspace. They repeated this process three times, adopting different postures on each occasion. As the robot arm approached the designated handover position, participants were able to retrieve objects provided by the robot. Throughout this process, the OpenPose algorithm was employed to create a skeleton model for each participant, capturing the movements of their upper arms, lower arms, and wrists.

Upon evaluating the handover gestures across 30 experimental trials involving 80% of the participants' skeleton models, it was observed that these models adhered to the RULA ergonomic metric. This outcome underscores the robustness of the robot's motion planning, which calculates interactive positions aligning with ergonomic requirements for human–robot handovers.

In the ergonomic test, instances where a minority of human skeleton models failed to meet the RULA guidelines can be attributed to two primary factors. Firstly, there were occasional visual estimation errors associated with the OpenPose algorithm in tracking human skeleton points. Secondly, variations in human movement patterns when approaching the designated positions introduced a degree of uncertainty.

As a concluding note, it is imperative to expand the evaluation of the prototype system for mutual-cognitive HRC by assessing its feasibility in diverse industrial applications, such as the assembly of complex mechanical engines.

4.4 Chapter summary

In this chapter, we delved into a approach that leverages MR technology and visual reasoning to enhance mutual-cognitive capabilities in Proactive HRC. The visual reasoning module was demonstrated in sequentially perceiving the on-site workspace, constructing an SG from these perceptual results, and subsequently developing task planning strategies through graph embedding. In the MR environment, this framework facilitated a bi-directional flow of information, where human operators received suggestions and support for their ongoing operations, while the robot obtained a understanding of the current scenario, enabling it to execute ergonomic actions.

To summarize, this chapter's scientific contributions can be outlined as follows:

a) Introduction of a visual reasoning approach into the HRC system, elevating its intelligence from a perception-based level to mutual-cognition. This reasoning module infers knowledge concerning human–robot collaborative workflows, facilitated by contextual SGs. It excels at inferring task-planning strategies across diverse cooperative scenarios.

b) Establishment of safety, preview, and ergonomics robot motions. It enhances the robot's empathetic capabilities. The system not only ensures the safety of human partners through visualized safety regulations but also provides trajectory previews within the MR environment. Additionally, the system excels in planning interactive positions that are within easy reach for human operators, thus reducing their physical stress.

References

Adeli, V., Adeli, E., Reid, I., Niebles, J.C., Rezatofighi, H., 2020. Socially and contextually aware human motion and pose forecasting. IEEE Robotics and Automation Letters 5, 6033–6040.

Bottani, E., Vignali, G., 2019. Augmented reality technology in the manufacturing industry: A review of the last decade. IISE Transactions 51, 284–310.

Cao, Z., Hidalgo Martinez, G., Simon, T., Wei, S., Sheikh, Y.A., 2019. OpenPose: Realtime multi-person 2D pose estimation using part affinity fields. IEEE Transactions on Pattern Analysis and Machine Intelligence.

Chen, Z., Chen, J., Geng, Y., Pan, J.Z., Yuan, Z., Chen, H., 2021. Zero-shot visual question answering using knowledge graph. In: The Semantic Web–ISWC 2021: 20th International Semantic Web Conference, ISWC 2021, Virtual Event, October 24–28, 2021, Proceedings 20. Springer, pp. 146–162.

Duberg, D., Jensfelt, P., 2020. UFOMap: An efficient probabilistic 3D mapping framework that embraces the unknown. IEEE Robotics and Automation Letters 5, 6411–6418.

He, K., Zhang, X., Ren, S., Sun, J., 2016. Deep residual learning for image recognition. In: Proceedings of the IEEE Conference on Computer Vision and Pattern Recognition, pp. 770–778.

Hernández, J.D., Sobti, S., Sciola, A., Moll, M., Kavraki, L.E., 2020. Increasing robot autonomy via motion planning and an augmented reality interface. IEEE Robotics and Automation Letters 5, 1017–1023.

Hietanen, A., Pieters, R., Lanz, M., Latokartano, J., Kämäräinen, J.K., 2020. AR-based interaction for human–robot collaborative manufacturing. Robotics and Computer-Integrated Manufacturing 63, 101891.

Kim, W., Lee, Y., 2019. Learning dynamics of attention: Human prior for interpretable machine reasoning. arXiv:1905.11666.

LaValle, S.M., 1998. Rapidly-exploring random trees: A new tool for path planning. The annual research report. https://api.semanticscholar.org/CorpusID:14744621.

Li, S., Zheng, P., Wang, Z., Fan, J., Wang, L., 2022. Dynamic scene graph for mutual-cognition generation in proactive human–robot collaboration. Procedia CIRP 107, 943–948.

Li, S., Zheng, P., Xia, L., Wang, X.V., Wang, L., 2023. Towards mutual-cognitive human–robot collaboration: A zero-shot visual reasoning method. In: 2023 IEEE 19th International Conference on Automation Science and Engineering (CASE). IEEE, pp. 1–6.

Li, S., Zheng, P., Zheng, L., 2020. An AR-assisted deep learning-based approach for automatic inspection of aviation connectors. IEEE Transactions on Industrial Informatics 17, 1721–1731.

McAtamney, L., Corlett, E.N., 1993. RULA: A survey method for the investigation of work-related upper limb disorders. Applied Ergonomics 24, 91–99.

Ong, S., Yew, A., Thanigaivel, N., Nee, A., 2020. Augmented reality-assisted robot programming system for industrial applications. Robotics and Computer-Integrated Manufacturing 61, 101820.

Rezatofighi, H., Tsoi, N., Gwak, J., Sadeghian, A., Reid, I., Savarese, S., 2019. Generalized intersection over union: A metric and a loss for bounding box regression. In: Proceedings of the IEEE/CVF Conference on Computer Vision and Pattern Recognition, pp. 658–666.

Shi, J., Zhang, H., Li, J., 2019. Explainable and explicit visual reasoning over scene graphs. In: Proceedings of the IEEE/CVF Conference on Computer Vision and Pattern Recognition, pp. 8376–8384.

Tang, K., Zhang, H., Wu, B., Luo, W., Liu, W., 2019. Learning to compose dynamic tree structures for visual contexts. In: Proceedings of the IEEE/CVF Conference on Computer Vision and Pattern Recognition, pp. 6619–6628.

Venkatesh, S.G., Biswas, A., Upadrashta, R., Srinivasan, V., Talukdar, P., Amrutur, B., 2020. Spatial reasoning from natural language instructions for robot manipulation. arXiv:2012.13693.

Wang, B., Zheng, P., Yin, Y., Shih, A., Wang, L., 2022. Toward human-centric smart manufacturing: A human–cyber-physical systems (HCPS) perspective. Journal of Manufacturing Systems.

Wang, L., 2022. A futuristic perspective on human-centric assembly. Journal of Manufacturing Systems 62, 199–201.

Yan, S., Xiong, Y., Lin, D., 2018. Spatial temporal graph convolutional networks for skeleton-based action recognition. In: Proceedings of the AAAI Conference on Artificial Intelligence.

Yang, J., Lu, J., Lee, S., Batra, D., Parikh, D., 2018. Graph R-CNN for scene graph generation. In: Proceedings of the European Conference on Computer Vision (ECCV), pp. 670–685.

CHAPTER 5

Predictable spatio-temporal collaboration

Industrial robots should possess the capability to discern human intentions by analyzing partially observed human actions and subsequently carry out assisting tasks. Nevertheless, conventional HRC in both industrial and academic settings concentrate on either reactive or adaptive robot planning. Such approaches involve waiting for human actions, potentially resulting in unanticipated safety concerns due to delayed responses. This impedes the seamless transition of HRC toward predictable teamwork. To address this bottleneck, we introduce an approach that leverages multimodal transfer learning to enable ongoing prediction of human intentions and proactive decision-making by robots. This approach aims to foster predictability in collaborative efforts in the near future. We evaluate the effectiveness of our method by demonstrating its application in an aircraft bracket assembly task.

5.1 Connotation

Rather than strict separation between pre-programmed industrial robots and shop-floor operators, HRC facilitates collaborative efforts where robots work alongside human operators in shared assembly tasks. For instance, Wang et al. (2020) conducted a review of HRC systems for welding processes, where robots have the ability to dynamically adjust their planned tasks to cooperate with human welders. However, these collaborative robots are often confined to a single monitored area and operate based on reactive instructions, falling short of efficiently integrating robotic automation and human cognition.

To address this challenge, predictable teamwork intelligence within Proactive HRC, is essential, which anticipates the execution of cooperative processes over time to optimize resource allocation. This enables a robot to predict human future intentions with only partial observation of human actions and make decisions that align with human expectations for performing manipulations.

For human intention prediction, a 3D Convolution Neural Network (CNN) (Wen et al., 2019) method was introduced to recognize opera-

Proactive Human–Robot Collaboration Toward Human-Centric Smart Manufacturing
https://doi.org/10.1016/B978-0-44-313943-7.00012-0

tors' actions during the assembly process using a visual controller. Yan et al. (2018) extracted action representations of In an HRC system, Liu and Wang (2021) inferred assembly context information from key points on the human body. However, these efforts often analyze human action intentions using a single modality in industrial settings, either through RGB video-based action sequences or skeleton-based ones. The former is effective at capturing fine-grained details but is rigid in associating visual patterns of the same action from different camera views, and it incurs high computing costs for long-duration videos. The latter lacks low-level detailed information. Thus, in real workplace settings for Proactive HRC, single-modality methods struggle to reliably predict operators' ongoing intentions due to challenges such as subtle and similar motion patterns within a short time frame, diverse visual patterns from different camera angles, and a shortage of labeled data for sequential actions of varying durations.

To advance predictable intelligence in Proactive HRC, several critical issues must be addressed, as depicted in Fig. 5.1. The first one is the acquisition of human sequential action data, including how to gather sufficient information within a short time series and reduce the annotation workload for extensive data collected in industrial settings. By combining visual features from videos and topology patterns from human skeleton joints, ample information about human action sequences can be obtained within a shorter time frame. Moreover, transfer learning algorithms can be applied to enable semi-supervised training, reducing the reliance on annotated data.

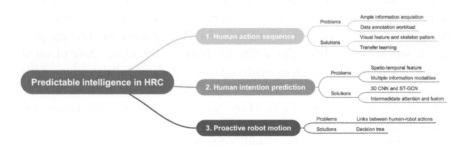

Figure 5.1 The connotation of predictable intelligence in Proactive HRC.

For human intention prediction, two key challenges need to be addressed: how to extract spatio-temporal features from both videos and human skeleton joints, and how to effectively integrate diverse information features from multiple modalities. In this context, the utilization of 3D CNN (Dang et al., 2020) and ST-GCN algorithms (Yan et al., 2018) allows for the extraction of video features and human topology patterns in spatio-temporal domains, respectively. Simultaneously, the incorporation of an intermediate attention mechanism and concatenation fusion layers facilitates the integration of these diverse features from observed human sequential actions.

Proactive robot motion enables a robot to plan its path in alignment with a human's intended goals in the immediate future. Decision tree algorithms offer a solution to establish a link between robotic planning and human intentions.

5.2 A multimodal human action prediction-based framework

The predictable HRC framework contains three steps, as visually depicted in Fig. 5.2. The first step involves the task of inferring and comprehending the complex human sequential actions. This phase leverages multimodal learning to transform the input data into multiple data modalities and distills feasible features in the temporal dimension.

Following this, the second step is the implementation of an online intention prediction algorithm. This algorithm predicts human intentions based on the partially observed data streams of human actions. An essential technology in this step is the utilization of a transfer learning algorithm. This strategy ensures the rapid and efficient deployment of the inference model, even in the face of scenarios where limited annotation data is available, spanning diverse industrial settings.

The final step is the dynamic decision-making mechanism in IIoT environments. In this context, decisions response to the foreseeable human intention give instructions to mobile robots. These instructions align with human expectations, ensuring collaboration as expected in the near future.

5.2.1 Multimodal intelligence-enabled human action pattern extraction

We present a multimodal intelligence-enabled approach for the prediction of real-time human actions, employing both RGB videos and skeletal joint data, as depicted in Fig. 5.3. The multimodal fusion network consists of three primary components. The first component utilizes an efficient inflated ResNet (Carreira and Zisserman, 2017) to discern visual action patterns. The second component leverages the ST-GCN algorithm (Yan

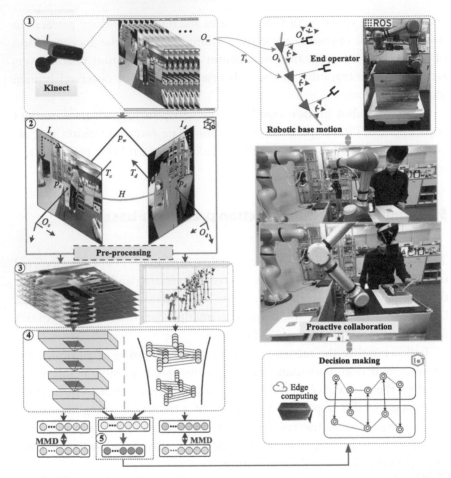

Figure 5.2 The overall framework of predictable intelligence in HRC. (Adapted from (Li et al., 2021b).)

et al., 2018) to perform element-wise maximization of action representations derived from the topological layout of the human body, utilizing information from skeletal joints. Lastly, the fusion component introduces an intermediate attention mechanism. This mechanism explores cross-channel relationships between the visual and skeletal modalities, enhancing the the feature fusion process (Su et al., 2020).

5.2.1.1 Efficient inflated ResNet for video representation classification

This research recognize visual action patterns by an Inflated ResNet50 module.

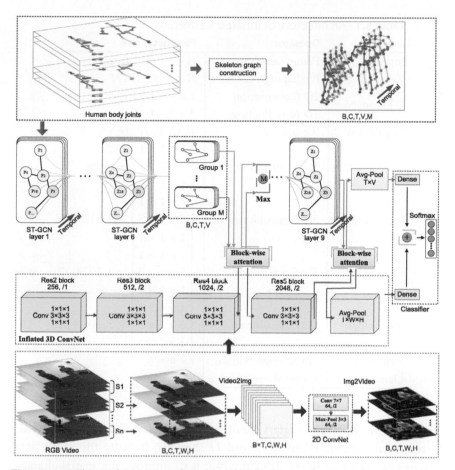

Figure 5.3 The architecture of multimodal fusion network for human action prediction. (Adapted from (Li et al., 2021b).)

In the visual stream, the input video undergoes a division into N equal sequences (i.e., $S_1, S_2, \ldots, S_n$), with one frame randomly selected from each partitioned sequence. This partitioning strategy enables the visual sub-network to effectively counteract instance variations within actions while accommodating a lightweight network architecture. This approach is vital for both long-term videos and high-frame-rate clips, which might contain redundant information.

To extract spatio-temporal features from the sampled frames, we employ an Inflated ResNet50 module, which is elaborated in Table 5.1. This module consists of both 2D ConvNet and inflated 3D ConvNet compo-

Table 5.1 Architecture of Inflated ResNet50.

Layer	Operator	Parameter size
Input	Frame sampling	B, C, T, W, H
Video2Img	Reshaping	$B \times T, C, W, H$
2D ConvNet	2D Convolution	$7 \times 7, 64, /2$
	2D Max Pooling	$3 \times 3, 64, /2$
Img2Video	Reshaping	B, C, T, W, H
Res2 block	3D Convolution	$\begin{bmatrix} 1 \times 1 \times 1 \\ 3 \times 3 \times 3 \\ 1 \times 1 \times 1 \end{bmatrix} \times 3, 256, /1$
Res3 block	3D Convolution	$\begin{bmatrix} 1 \times 1 \times 1 \\ 3 \times 3 \times 3 \\ 1 \times 1 \times 1 \end{bmatrix} \times 3, 512, /2$
Res4 block	3D Convolution	$\begin{bmatrix} 1 \times 1 \times 1 \\ 3 \times 3 \times 3 \\ 1 \times 1 \times 1 \end{bmatrix} \times 3, 1024, /2$
Res5 block	3D Convolution	$\begin{bmatrix} 1 \times 1 \times 1 \\ 3 \times 3 \times 3 \\ 1 \times 1 \times 1 \end{bmatrix} \times 3, 2048, /2$
Avg-Pool	3D Average Pooling	$t \times w \times h$
Dense	Linear Regression	K

nents. The network parameters, denoted as B (batch size), C (channels), T (temporal length), W (width), and H (height), indicate the input data dimensions. The role of the 2D ConvNet is to improve the spatial characteristics of each RGB frame across the temporal dimension. Following this enhancement, the inflated 3D ConvNet directly delves into the exploration of spatio-temporal representations within these improved features.

The transformation of 2D filters from $N \times N$ to cubic 3D filters of $N \times N \times N$ achieves a crucial expansion for convolutional and pooling filters, allowing them to incorporate temporal modeling characteristics. These 3D kernels are initialized by replicating the pretrained weights of the 2D filters N times along the temporal dimension, and to ensure consistent scaling, they are normalized by dividing by N. This approach effectively eliminates the need to train the 3D ConvNet from the scratch, thus mitigating the challenges associated with an increased number of training parameters. As a final step in the process, a dense layer is appended to the 3D ConvNet, enhancing its capacity to predict the classification of K distinct action classes.

5.2.1.2 Element-wise ST-GCN for human body topology extraction

The skeleton stream includes nine ST–GCN layers, and an element–wise maximization mechanism. The process is demonstrated in Fig. 5.4. In assembly workplaces, Azure Kinect is used to capture the motions of operators during ongoing tasks. Subsequently, the captured motion data is processed to derive human pose information. This is achieved through the utilization of the Openpose toolbox (Cao et al., 2017), which can estimate the precise locations of the operator's body joints across consecutive frames of a video. Through the tracking of body joints, the ST–GCN model can extract human body topology information.

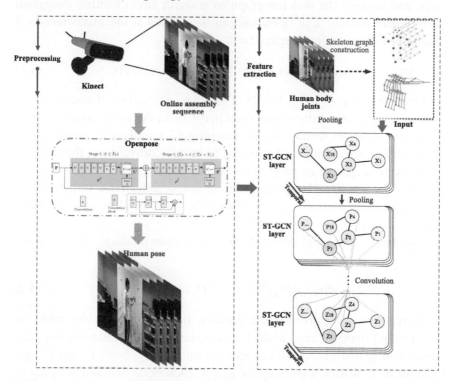

Figure 5.4 The ST-GCN model for feature extraction of human body topologies. (Adapted from (Li et al., 2021a).)

In order to facilitate the initiation of the ST-GCN layers and to provide the input they require, the first step involves the conversion of human skeleton sequences into an undirected spatio-temporal graph, which is denoted as $G = (V, E)$. This graph construction is conducted through the following process: The node set, represented as $V = \{v_{ti}|t = 1, \ldots, T; i = 1, \ldots, N\}$, reflecting the video in T frames, with each frame containing N distinct human joints. Moreover, the edge set E exhibits comprise two unique subsets. The first subset is the intra–skeleton connection $E_S = \{v_{ti}v_{tj}|(i,j) \in H\}$, which defines the intrinsic links and associations between joints within each frame. It captures the inherent and organic connections that exist among various human body joints within a single frame. The second subset is inter-frame edges $E_F = \{v_{ti}v_{(t+1)i}\}$ across consecutive frames. These inter-frame edges link identical joints across different frames and thereby encapsulating the dynamic characteristics of the human skeleton over time.

Subsequently, graph convolution filters are deployed in the ST-GCN layer, and traverse the skeleton graph to unearth latent features associated with human actions. This operation can be defined as the spatio-temporal graph convolution, characterized by

$$f_{out} = \Lambda^{-\frac{1}{2}}(A + I)\Lambda^{-\frac{1}{2}}f_{in}(P(v_{ti})) \cdot w(l_{st}(v_{ti})). \tag{5.1}$$

Here, the term $\lambda^{ii} = \sum_j A^{ij} + I^{ij}$. The adjacency matrix A and the identity matrix I represent the connections between nodes in the graph G. The remaining elements involve feature mapping f_{in}, sampling P, weighting w, and partition operations l_{st}. Analogous to image convolution, the sampled nodes from the graph are projected into a c-dimensional space through the use of $f_{in} : V^2 \rightarrow \mathbb{R}^c$. The weighting function w generates a weight vector of identical dimensionality c to compute the inner product between these nodes.

To identify the neighboring nodes for a given node v_{ti}, the sampling function P is employed:

$$P(v_{ti}) = \{v_{qj}|d(v_{tj}, v_{ti}) \leq D, |q - t| \leq \Gamma\}. \tag{5.2}$$

Here, $d(v_{tj}, v_{ti})$ represents the distance between two nodes within a graph frame denoted as t, while Γ corresponds to the temporal span between two spatial graphs. In our experimental setup, both D and Γ have been set to 1. Within a given spatial graph, each node denoted as v_{tj} has its neighboring nodes categorized into three distinct subsets using a partition strategy denoted as l. This partitioning process is facilitated by calculating the gravity center v_c for the respective spatial graph and can be articulated as follows:

$$l_{ti}(v_{tj}) = \begin{cases} 0, & \text{if } d(v_{tj}, v_c) = d(v_{ti}, v_c), \\ 1, & \text{if } d(v_{tj}, v_c) < d(v_{ti}, v_c), \\ 2, & \text{if } d(v_{tj}, v_c) > d(v_{ti}, v_c). \end{cases} \tag{5.3}$$

In a similar way, the neighboring node v_{qj} associated with node v_{ti} is denoted as l_{st} within the spatio–temporal graph. As the graph convolution filters traverse the graph, they assign varying weights to these nodes, thereby capturing both spatial and temporal relationships:

$$l_{st}(v_{tj}) = l_{qi}(v_{qj}) + (q - t + \Gamma) \times 3. \tag{5.4}$$

In this approach, a stack of six ST-GCN layers is applied to the graph G, facilitating the distillation of low-level information of human actions, such as joint motion. Following this, the extracted feature maps are divided into M groups, with M representing the number of individuals in the input data. To capture high-level skeleton patterns associated with human actions, an element-wise maximization mechanism is employed. This mechanism compares elements within the feature maps across groups, selecting the maximum value from each comparison. These maximum values are subsequently fed into an additional set of three ST-GCN layers, enhancing the extraction of high-level patterns in the skeletal data. The final steps involve pooling operations and linear regression for the construction of the action classifier.

5.2.1.3 Intermediate attention and fusion for multi-modalities

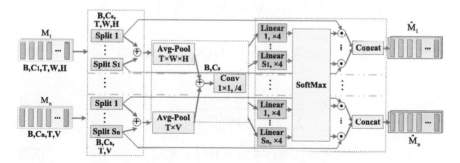

Figure 5.5 The procedure of the intermediate attention module. (Adapted from (Li et al., 2021b).)

The integration of the visual and skeleton modalities is achieved through a combination of intermediate attention and late fusion mechanism, as depicted in Fig. 5.5. The intermediate attention process optimizes feature maps, denoted as $\hat{M}_1, \ldots, \hat{M}_n$, by exploring inter-contextual relationships among input modalities, represented as $M_1, \ldots, M_n$. This optimization involves three steps:

a) (Segmentation) Feature maps from modality M_1 to M_n are divided into equal blocks, denoted as $S_i, i \in 1, \ldots, n$. The number of blocks in each modality M_i is determined by $|S_i| = \lceil C_i/C_s \rceil$, where C_i represents the number of channels in M_i. In cases where C_i is not a multiple of $C_s = \min[S_1, \ldots, S_n]/2$, the last block in S_i is padded with zeros. Blocks within a modality M_i are then related to each other through element-wise summation over $S_1, \ldots, S_i$.

b) (Connection) The output of the segmentation process is represented as D_i within modality M_i, where $D_i \in \mathbb{R}^{N_1 \times \cdots \times N_k \times C_s}$. After applying global average pooling P_i to D_i, as described in Eq. (5.5), multimodal contextual information is aggregated by summing $P_1, \ldots, P_n$. Subsequently, a 1×1 CNN layer is introduced to capture cross-channel relationships among these multimodal features.

c) (Activation) The connection phase yields a global shared representation, denoted as G. Attention weights W_i^j for the j-th block of the i-th modality are generated through sequential linear transformations and SoftMax activations on G, as detailed in Eq. (5.6), with parameters $i \in 1, \ldots, n$ and $j \in 1, \ldots, |S_i|$. Subsequently, an optimized feature block $\hat{S}_i^j$ is obtained using Eq. (5.7). The parameter λ is set to 0.5 in our algorithm. Finally, the optimized output $\hat{M}_i$ is produced through concatenation with $\hat{S}_i^j$, i.e., $\hat{M}_i = \left[\hat{S}_i^1, \ldots, \hat{S}_i^{|S_i|} \right]$.

$$P_i(C_s) = \frac{1}{\prod_{j=1}^{k} (N_1, \ldots, N_k)} \sum_{(N_1, \ldots, N_k)} D_i(N_1, \ldots, N_k, C_s), \tag{5.5}$$

$$W_i^j = \frac{\exp(h_i^j G + b_i^j)}{\sum_i^n \sum_j^{|S_i|} \exp(h_i^j G + b_i^j)}, \tag{5.6}$$

$$\hat{S}_i^j = \left[\lambda + (1 - \lambda) \times W_i^j \right] \times S_i^j. \tag{5.7}$$

In Fig. 5.3, the model architecture incorporates two intermediate attention mechanisms, which can enhance the fusion of visual and skeleton modalities. The first intermediate attention module is nestled between the fourth layer of the inflated ResNet50 and the person-wise groupings in the sixth ST-GCN layer. Simultaneously, the second intermediate attention module takes its place between the fifth layer of the visual stream and the ninth ST-GCN layer. These modules act as block-wise correlations, fusing the visual and skeleton representations. This fusion extends from the shallow feature recalibration to the acquisition of high-level contextual awareness.

Then, the outputs of the visual and skeleton part embark on a convergence through element-wise addition. For action pattern classification, a SoftMax layer is applied. During the model's training phase, the cross-entropy loss function is leveraged to optimize the model's weights. These weights undergo adjustment, aligning them with the desired performance objectives and ensuring that the model can predict action patterns with accuracy and precision.

5.2.2 Transfer learning-based online human intention prediction

This approach also introduces a transfer learning module, aimed at the online prediction of human intentions from partially observed human action sequences in practical scenarios. This process, outlined in Fig. 5.2 (steps ❶ to ❺), addresses two challenges:

a) (Knowledge Transfer for Pattern Recognition) One of challenges in intention prediction is the transfer of knowledge of pattern representation of human sequential actions with limited annotation data.

b) (Online Human Intention Prediction) The second challenge involves the real-time prediction of human intentions based on partially observed human action sequences.

These challenges are tackled in the subsequent sections.

5.2.2.1 Pattern recognition of human sequential action with few data

The process of annotating data in various scenarios within extensive sensor networks in modern factories is known to be both costly and time-consuming. To address this challenge, a semi-supervised transfer learning approach is introduced. This approach entails the following steps:

a) (Fine-Tuning Strategy) Firstly, a fine-tuning strategy is employed to enable the extractor G to acquire knowledge of action patterns, starting from daily activities and extending to industrial actions.

b) (Domain Adaptation Module) A domain adaptation module denoted as θ_m is integrated with G. In both the visual and skeleton streams, represented as θ_m^v and θ_m^s respectively, this module serves to mitigate the distribution differences between source and target data. This alignment process facilitates G in aligning learned action representations derived from a limited set of labeled data with the remaining unlabeled data.

c) (Feature Mapping) To achieve this alignment, a fully connected layer is applied to map features from both the source domain, denoted as δ_s, and the target domain, denoted as δ_t. The distribution distance between these two domains is subsequently computed using the Maximum Mean Discrepancy (MMD) metric.

These steps collectively contribute to the development of the semisupervised transfer learning procedure, which covers the challenges associated with data annotation in diverse industrial contexts. We consider:

$$D = \|\delta_s - \delta_t\|_H = \|\langle \delta_s, \delta_t \rangle_H\|. \tag{5.8}$$

Here, the notation H represents the reproducing kernel Hilbert space. The unbiased estimation value for $\langle \delta_s, \delta_t \rangle_H$ is determined using the Gaussian radial basis function, specifically, $\hat{D} = \exp(\|\delta_s - \delta_t\|^2 / 2\sigma)$.

5.2.2.2 Online human intention prediction ahead of time

The process of online human intention prediction from partially observed human action sequences is delineated into five sequential steps, illustrated in Fig. 5.2: ❶ RGB-D video acquisition, which captures the primary data source; ❷ color–depth camera calibration, which is carried out to align the color and depth cameras, ensuring accurate spatial correspondence between the two modalities; ❸ visual and skeleton output, which enables the generation of synchronized visual and skeleton data from the captured RGB-D videos; ❹ feature extraction, which is applied to the synchronized data to distill relevant information for subsequent analysis; and ❺ pattern classification, which is employed to discern human intentions based on the extracted features.

In workshop settings, the Azure Kinect system is commonly employed for recording RGB-D videos and extracting 3D skeleton joint information. However, it is worth noting that the human subject typically occupies only a small portion of the video frames. To mitigate background interference, various pre-processing strategies are applied to enhance data quality.

To convert the 3D joint data into 2D pixel coordinates, a calibration process is initiated. This aligns the color camera and the depth camera within the Kinect system. The transformation matrix from the depth camera to the color image is computed based on the camera model $P_{uv} = TP_c = KTP_w$, ensuring accurate spatial mapping between the depth and color modalities:

$$P_{uv-c} = K_c T_c T_d^{-1} P_d. \tag{5.9}$$

Here, the transformation of 3D joint data into 2D pixel coordinates relies on the calibration of both the color camera (parameterized as c) and the depth camera (parameterized as d). The camera intrinsics, denoted as K, and camera extrinsics, represented by $T = [R|t]$, are used in this calibration process. As a result, precise 2D body joint coordinates are obtained. Additionally, the dimensions of a bounding box surrounding the person in the scene, namely its width w_b and height h_b can be determined. To crop the human part from images uniformly, a crop mask with a fixed aspect ratio α is introduced, where w_m is calculated as $w_m = \max\{\lfloor w_b, h_b/2 \rfloor\} + d_h$, with d_h representing the distance from the pelvis to the middle of the spine. In this work, α is set to 2.

For skeleton data, a normalization process proposed by (Shahroudy et al., 2016) is applied. Leveraging both visual and skeleton output, a pretrained multimodal model is employed to estimate human operational intentions.

Simultaneously, when dealing with live data streams, the detailed procedure for online intention analysis is presented in Algorithm 1. A queue denoted as Q, capable of holding N frames, is employed to record the incoming video stream. Additionally, a memory queue M is utilized to store T temporal video segments, each of which matches the size of Q. Here, N frames represent a timestamp, indicating the number of input frames for the action prediction model. Upon initiation of the live video stream, all frames are uniformly transferred to M until it reaches capacity. Subsequently, half of the frames from Q are enqueued into M to prevent overflow.

For action prediction, input data is sampled from these T temporal segments within M. Assuming $T = 3$, I_d encompasses 25% of the samples from Q at the current time step M_0, 25% samples from Q at M_1, and 50% samples from Q at the last timestamp M_2. This online mode effectively captures long-range information from the incoming stream, while assigning greater significance to recent frames. Hence, the online mode can anticipate human intentions across varying time spans.

5.2.3 Proactive robot motion planning

This section elaborates the proactive robot motion planning in HRC, including robot perception of surrounding environments, the decision-making process, and system deployment structure.

Algorithm 1 Online human intention prediction.

Input: RGB-D live video stream (V_d)

Number of visual-stream input (N)

Preprocessing module (Δ)

Pretrained action pattern recognition model (Φ)

Output: Human intention predictions

1: Calculate the number of timestamps $T = \lceil \log_{0.5} \frac{1}{N} \rceil$
2: Initialize video queue Q for N coming frames
3: Initialize memory queue M for $N \times T$ frames
4: Mark M with timestamps, i.e., $M = \{M_0, \dots, M_T\}$
5: **while** new frames available from V_d **do**
6:     Add RGB-D frame f_i from V_d to queue Q
7:     **if** $i\%N$ **then**
8:     **else if** $i < N \times T$ **then**
9:         Add N frames from Q to queue M
10:        $I_d := \{0.5^{\frac{1}{N}} M_0\} \bigcup_{t=1}^{N} \{0.5^{\frac{1}{N}-t+1} M_t\}$
11:        Add 50% frames from Q to M and update
12:        $I_d := \{0.5^{T} M_0\} \bigcup_{t=1}^{T} \{0.5^{T-t+1} M_t\}$
13:        Feed I_d to Δ and Φ to obtain prediction

5.2.3.1 Robot perception of the surrounding environment

In addition to observing human actions, the generation of decisions relies heavily on the robot's capacity to perceive its surrounding environment. This ability to understand and interpret the environment is an essential prerequisite, which contains 6-DoF pose estimation of industrial parts and 3D point cloud segmentation of the working environment.

A 6-DoF pose estimation is introduced to recognize the 3D location and posture of target workpieces, answering the questions of "where" and "what" in relation to industrial components for robot operations. The upper segment of Fig. 5.6 is the high-resolution network-based approach. Remarkably, it showcases a robust capability in handling instances where occlusions occur due to the presence of human hands or robot grippers, a common occurrence in the dynamic HRC environment.

Firstly, we introduce a modified HRNet (Fan et al., 2021) into HRC system to extract high-resolution features from observed visual images. Notably, this modified HRNet incorporates a mask-guided attention mechanism, which can model and identify occlusion areas within the images.

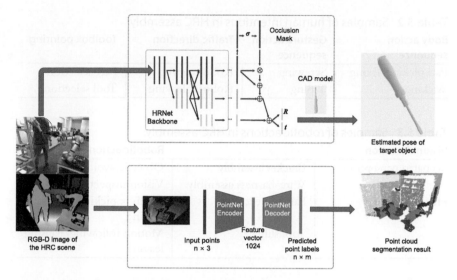

Figure 5.6 The robot perception of object 6-DoF pose and 3D working environment. (Adapted from (Zheng et al., 2022).)

This mechanism ensures that the feature map extracted from the visual data is not only high-resolution but also takes into account the presence of occluded regions. This model needs to predict two sets of parameters: the translation parameters denoted as $t = (t_x, t_y, t_z)^T$ and the rotation parameters represented as $R = (R_{roll}, R_{pitch}, R_{yaw})^T$. These parameters are instrumental in precisely defining the workpiece's posture. During the training process, the model is optimized using a loss function that let the predicted pose parameters as close as possible to the corresponding ground truth parameters calculated from the 3D CAD model. This training process ensures that the model can estimate the workpiece's posture with a high degree of accuracy.

Additionally, the parsing of the working environment empowers the robot to comprehend and construct a geometric representation of the entire workspace. In this context, the 3D point cloud segmentation method can be used, as illustrated in the lower section of Fig. 5.6.

To begin the process, the acquired visual and depth images, essentially the RGB-D data, undergo an initial transformation into a point cloud format. This resulting point cloud is structured as an array with dimensions $n \times 3$, where each point within it is defined by a set of 3D coordinates. Here, n symbolizes the total count of such points contained within the cloud. The subsequent step involves the application of the PointNet model to these point clouds. This model is employed for feature extraction of the

Table 5.2 Samples of human intentions in HRC assembly.

Body action sequence	Gesture action sequence	Traffic direction	Toolbox pointing
Part picking/fixing	Screwing	Robotic guiding	Part selection
Walking	Taping	Robotic leaving	Tool selection

Table 5.3 Samples of robotic actions in HRC assembly.

Vision detection		Robotic action
Scenario perception	Bracket assembly	Obstacle avoidance
	Wire-harness assembly	Vision inspection
Human holding	Part subject	Toolbox picking/holding/placing
	Tool subject	Motion following/pausing/leaving

3D point cloud. Consequently, the original features present in the point cloud are encoded into vectors, each of which holds 1024 dimensions. As the feature extraction proceeds, the encoded vectors are then decoded in a sequential manner. This decoding process eventually generates a matrix sized $n \times m$. In this matrix, m pertains to the various categories that are assigned to each individual point within the point cloud. The final step involves the application of a softmax operation to the resulting matrix. This operation generates the ultimate point cloud segmentation results. Within these results, each point is precisely labeled with its respective category, thereby providing a clear and concise description of the entire point cloud.

5.2.3.2 Dynamic decision-making mechanism

Building upon the earlier predictions of human intentions and environmental perception, a dynamic decision-making mechanism is incorporated to enable proactive robotic planning that aligns with human expectations. Employing the decision-making module, a mobile robot gains insight into anticipated human intentions, thereby facilitating intelligent assistance to the operator. Several examples illustrating human intentions and corresponding robotic actions within a human-centered assembly environment are provided in Tables 5.2 and 5.3 for reference.

To elucidate the decision-making process, Table 5.4 outlines the associations between human intentions and corresponding robot actions. Specifically, "placing" signifies that robots are tasked with delivering a tool-

Table 5.4 An example of industrial HRC.

Number	Traffic direction	Body action sequence	Gesture action sequence	Toolbox pointing	Robotic decision
1	Guiding	Picking	Taping	Part	Part placing
2	Guiding	Picking	Taping	Tool	Tool placing
3	Guiding	Fixing	Screwing	Part	Part picking
4	Guiding	Fixing	Screwing	Tool	Tool picking
5	Leaving	Fixing	Screwing	Part	Part picking
6	Leaving	Picking	Taping	Part	Part picking
7	Leaving	Picking	Taping	Tool	Tool picking
8	Leaving	Fixing	Screwing	Tool	Tool picking
9	Guiding	Walking	Taping	Part	Part holding
10	Guiding	Walking	Taping	Tool	Tool holding

box to the human operator in close proximity, while "picking" implies that robots leave the operator's side to retrieve a toolbox from designated storage areas. In response to the "holding" instruction, the robot maintains close proximity to the human while carrying a toolbox.

The decision-making process utilizes the Iterative Dichotomiser 3 (ID3) algorithm to construct a multiway tree, leveraging the attribute with the information gain among the decision categories, which pertain to various classes of robotic actions observed in our samples. The information learned from this decision-making process is quantified as entropy,

$$\text{Ent}(D) = - \sum_{k=1}^{|y|} p_k \log_2 p_k. \tag{5.10}$$

Here, we have $|y|$ representing the number of decision categories, which in our case is equal to six. Additionally, P_k denotes the proportion of the k-th decision category.

The resulting decision tree, as depicted in Fig. 5.7, serves as a general framework connecting human intentions to robotic planning. Within this tree structure, the robotic decision at the final leaf node is generated and serves as feedback to guide human actions throughout the collaborative process, facilitating seamless coordination from top to bottom.

In this way, this approach enables the realization of flexible and predictable decisions in HRC assembly tasks, with proactive robotic motions aligning with human expectations in efficient co-working environment.

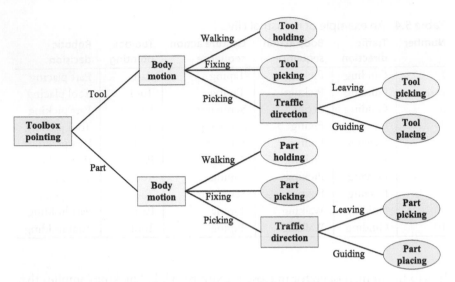

Figure 5.7 Decision tree for semantic knowledge linking between robot and the surrounding observation. (Adapted from (Li et al., 2021a).)

5.2.3.3 System deployment structure

The implementation of the predictable HRC system is in the IIoT environment. This setup is characterized by high-performance computing resources in the cloud plane and an array of hardware and software components situated in the edge plane. The collaborative interplay of these components is represented in Fig. 5.8. These components include Kinect, HoloLens, ROS, a robot controller, and a GPU server, forming a comprehensive system that allows seamless data communication and analysis. Devices in the edge plane are used for real-time processing of visual data, like the Kinect sensor. It captures and feeds real-time visual information into the system. Furthermore, the MR device, HoloLens, is introduced to enhance the operator's experience and interaction with the robotic system. By overlaying digital information onto the physical workspace, HoloLens provides interface for data visualization in collaboration. Meanwhile, the ROS-based software framework provides the communication and coordination channel between the various hardware components, ensuring that data and commands flow seamlessly. The robot controller acts as the bridge between the cloud-based computing and the local edge components for robotic operations. It translates high-level decisions and commands into precise robot motions, ensuring that HRC is smooth and safe. Adding to the system's processing capabilities is the GPU server, a powerful computational resource that handles data-intensive tasks. It enables rapid analysis and decision-making based on the data captured by the Kinect. The synergy between these edge devices and the cloud-based computing infrastructure transforms raw visual data into actionable knowledge, facilitating collaborative operations between human operators and robotic systems.

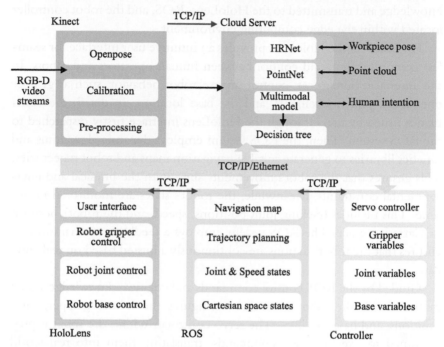

Figure 5.8 The system deployment structure in an IIoT environment. (Adapted from (Zheng et al., 2022).)

In edge computing, Openpose is used to extract human skeleton joints. Beyond this, the calibration processes including the color-depth cameras and robot hand–eye calibration are completed. The calibration procedure establishes a precise connection between visual sensors and robot kinematics. The data streams, comprising both RGB and depth information, undergo a pre-processing procedure. This phase segregates RGB-D video streams into visual and depth counterparts. Then, the data is dispatched to the cloud server via the TCP/IP protocol, for further proceeding. In the cloud server, two deep learning models, i.e., pretrained HRNet and PointNet, is deployed to estimate the 6D poses of workpieces and generate the point cloud representations of the environment. These perceptual results can be leveraged to guide robotic operations within the shared HRC workspace. On the other hand, the multimodal human action prediction model is used to forecast human intentions in advance. Based on the prediction, the system formulates reasonable collaboration strategies, encompassing cooperative work arrangements and proactive robot motions. Subsequently, all decisions regarding robotic motion are stored as retrievable knowledge and transmitted to the HoloLens, ROS, and the robot controller located within the edge computing environment.

For instance, the HoloLens provides an intuitive user interface for seamless communication and control between human operators and robots. In this interface, human can control diverse robot behavior, such as gripper operations, joint movements, and the base location. In detail, the commands issued by users through the HoloLens interface firstly dispatched to the ROS system. Then, the ROS system employs dynamic algorithms and kinetics libraries to generate suitable navigation maps and robot trajectories. This ensures that the robot's movements align with the physical and environmental constraints. Meanwhile, the robot's states are monitored in real time. This includes tracking joint positions, speed, and the robot's position in Cartesian space. The ROS system employs a feedback mechanism that ensures that the robot's actions are continuously adjusted based on real-time data.

Guided by the ROS control commands and cognitive knowledge stored in the cloud, the robot controller configures the robot's gripper, joint positions, and base variables. The servo controller within the robot is programmed to execute these commands, translating them into real-world robot actions. The robotic base, the underpinning of the robot's physical movements, is described by the equation $O_b^i = T_b^i O_w$, $i = p_1, \ldots, p_n$, with i representing different locations the robot base can traverse. Similarly, the robot's end effector follows a distinct trajectory denoted as $\{O_{e_j}^{p_1}, \ldots, O_{e_j}^{p_n}\}$, where j represents the various actions carried out by the robot controllers. With this framework of real-time robot states, the robot controller can dynamically adjust contact forces and robot positions. Hence, precise and safe robotic control during collaborative work with human operators in close proximity is achieved.

5.3 Case study

This section presents a demonstrative case study focused on assessing the predictability of Proactive HRC through bracket assembly tasks within an aircraft cabin laboratory environment. The objective is to further evaluate the effectiveness of our proposed approach.

Within the aviation industry, assembly tasks still rely on manual labor coupled with domain expertise, due to the confined workspace within aircraft cabins. During the assembly process, operators are frequently required to shuttle between the assembly area and a tool storage location to exchange toolboxes, resulting in a heavy and monotonous workload with suboptimal efficiency.

To address this challenge, we introduce a mobile robot to handle pick-and-place tasks within a human–centered assembly setting. As depicted in Fig. 5.9, the HRC framework for bracket assembly is realized by integrating online human intention prediction and proactive robotic decision-making. This system allows predictable co-working between human and robotic agents, thereby improving overall task efficiency and reducing the burdensome nature of the assembly process.

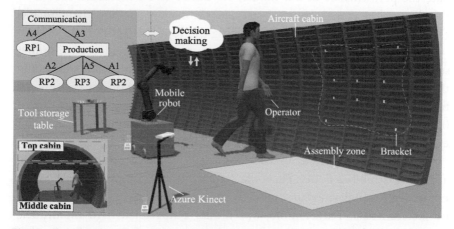

Figure 5.9 Application of predictable HRC in the task of aircraft bracket assembly. (Adapted from (Li et al., 2021b).)

5.3.1 Human action pattern recognition and intention prediction

The proposed multimodal transfer learning-based network plays a pivotal role in the recognition of human action patterns and the prediction of human intentions, which is the foundation of predictable HRC. In our experimental setup, an open human action dataset is leveraged to assess the effectiveness of the proposed multimodal fusion network. Besides, we

Table 5.5 Action prediction performance on the NTU-RGB+D dataset.

Model	X-Sub	X-View
ST-GCN (Skeleton only) (Yan et al., 2018)	81.50%	88.30%
Glimpse Clouds (RGB only) (Baradel et al., 2018)	89.60%	93.20%
RGB-Skeleton fusion (Liu et al., 2019)	85.40%	91.60%
IR-Skeleton fusion (De Boissiere and Noumeir, 2020)	91.80%	94.90%
Ours (RGB-Skeleton)	91.10%	96.00%

proceed to develop an Assembly Action Dataset (AAD) to showcase the knowledge transfer performance in predicting human intentions.

5.3.1.1 Evaluation of numerous action pattern recognition

The NTU-RGB+D dataset, an openly available resource, encompasses a diverse range of 60 daily action classes (Shahroudy et al., 2016). To evaluate the performance of our model, we utilize two benchmark subsets of this dataset: the cross-subject (X-sub) and cross-view (X-view) benchmarks (Yan et al., 2018), focusing on measuring the top-1 recognition accuracy.

Training. In our training process, we employ the video partitioning strategy that samples 15 RGB frames for the visual stream, while initializing the skeleton stream with 300 frames. The optimization of our multimodal fusion model is the SGD algorithm, starting with an initial learning rate of 0.001. This learning rate is subsequently reduced by a factor of 0.1 after every 10 epochs to facilitate convergence.

Results. As depicted in Table 5.5, our proposed multimodal fusion approach exhibits a noticeable enhancement in recognition accuracy when compared to single-modality models, such as those relying solely on RGB or skeleton data. Furthermore, our model outperforms two multimodal fusion methods. In a similar vein to the concept of Frames Per Second (FPS), our approach's speed is quantified in terms of Videos Per Second (VPS). When executed on a Tesla V100 GPU with 16 GB of memory, our model achieves a processing rate of 17 VPS, meeting the real-time requirements for online action prediction.

5.3.1.2 Evaluation of intention prediction with fewer frames

Fig. 5.10 illustrates the samples of AAD, which comprises sequential actions performed by operators during bracket assembly tasks. In this scenario, operators may carry tools and utilize smart equipment, such as MR glasses, to collaborate effectively with a mobile robot. The AAD includes a total of

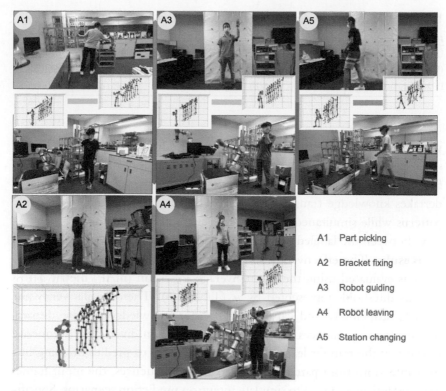

Figure 5.10 Examples of operators' action sequences in assembly tasks. (Adapted from (Li et al., 2021b).)

256 RGB-D videos captured using Azure Kinect in five operation intentions, denoted as A1) part picking, A2) bracket fixing, A3) robot guiding, A4) robot leaving, and A5) station changing. Importantly, the durations of operator actions tend to vary and are uncertain. These action sequences span from two seconds to five seconds within an unequal time length. On average, there are three to five action groups within a single video clip, resulting in a dataset containing a maximum of 939 action samples. Each RGB-D video includes visual frames at a resolution of 640×576 pixels and 3D skeleton poses, encompassing 25 body joints.

Experimental setup. The AAD is divided into two sets: a training set comprising 467 samples and a testing set comprising 472 samples. To simulate real-world scenarios within modern factories, the training dataset introduces a controlled variable μ, which represents the percentage of annotated data. This simulation is akin to situations where data collected

Table 5.6 Intention prediction performance on the AAD dataset.

μ	A1	A2	A3	A4	A5	mAP
0.05	91.11%	98.55%	92.59%	99.26%	94.23%	95.15%
0.10	97.04%	100.00%	100.00%	100.00%	98.08%	99.02%
0.20	100.00%	100.00%	100.00%	100.00%	100.00%	100.00%

via extensive sensor networks lacks labeling. Additionally, the optimization settings remain consistent with the training procedure of the multimodal fusion model. During the training process, our semi-supervised model undertakes knowledge transfer from the NTU dataset to adapt to assembly patterns while simultaneously extracting shared and domain-invariant features between the labeled target domain and the unlabeled source domain.

Results. The effectiveness of our approach's knowledge transfer capability is evaluated using the mean Average Precision (mAP) metric across various thresholds represented by μ (Table 5.6). These results verify the model's adaptability and capacity to transfer knowledge despite disparities in motion sequences and visual patterns within diverse environments. Given that the transfer learning-based model aims to predict ongoing human intentions from partially observed action sequences, the input frames are substantially reduced to simulate intention prediction scenarios. Specifically, the visual stream employs 15 RGB frames as input, while the skeleton subnet samples 50 frames. Remarkably, the model can forecast operators' intentions ahead of time by analyzing the first portion of the videos and generating predictions. As depicted in Fig. 5.11, the proposed model is capable of delivering early intention predictions in advance of 30% of the timestamps for action video streams. This predictive capability holds practical significance in real industrial settings where different workers exhibit varying durations for the same operation. Despite the uncertainty in execution time, the model effectively anticipates the operator's intention by analyzing a short segment of the action sequences.

5.3.2 Predictable teamwork in HRC for aircraft bracket assembly

Based on human intention prediction and analysis in HRC, predictable teamwork in the near future can be achieved by employing robotic dynamic decisions and proactive motion control.

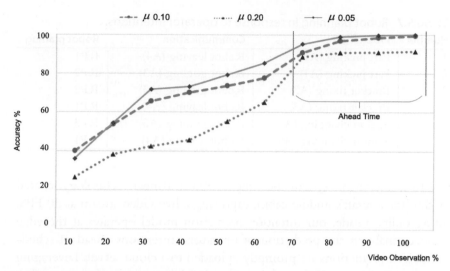

Figure 5.11 Early action prediction results in the AAD dataset. (Adapted from (Li et al., 2021b).)

5.3.2.1 Robotic dynamic decision-making

Within the context of human–centered aircraft bracket assembly tasks, the mobile robot can provide predictable assistance to the operator in various scenarios. Table 5.7 outlines six distinct cooperation scenarios that may occur during the production and communication process. In response to human actions, the robot offers three modes of assistance, which are as follows:

a) Moving away from the operator to retrieve toolboxes from storage areas (RP1).

b) Approaching the operator to deliver toolboxes (RP2).

c) Tracking the operator's movements to aid in holding tools or parts (RP3).

To acquire the knowledge required to establish these semantic links for cooperation, we employ the ID3 algorithm, which generates three potential decisions for robot planning based on observed human actions and intentions.

5.3.2.2 System deployment

Within the IIoT environment, a demonstrative system has been deployed for achieving predictable intelligence in HRC, particularly in bracket assembly tasks, as depicted in Fig. 5.9.

Table 5.7 Robot planning in response to operator intentions.

Number	Production	Communication	Robot planing
1	Part picking (A1)	Robot leaving (A4)	RP1
2	Part picking (A1)	Robot guiding (A3)	RP2
3	Bracket fixing (A3)	Robot guiding (A3)	RP2
4	Bracket fixing (A3)	Robot leaving (A4)	RP1
5	Station changing (A5)	Robot guiding (A3)	RP3
6	Station changing (A5)	Robot leaving (A4)	RP1

To facilitate this system, we employ Azure Kinect technology located within the aircraft's middle cabin, capturing a live video stream at 30 FPS. In an online mode, our intention prediction model operates at the edge service, enabling the prediction of operators' intentions ahead of schedule. These predictions are promptly uploaded to a cloud server. Leveraging these predictions, an ID3 module in the cloud dynamically generates robot decisions in real-time.

The mobile robot, composed of a Universal Robots UR5 integrated with an AGV base, receives these generated robotic control instructions in advance, which includes programming for position and speed. The instructions allow the robot to proactively assist the operator by anticipating their upcoming operation objectives. In this way, the integration of predictable intelligence into Proactive HRC allows high-level collaboration and enhanced productivity within the bracket assembly task.

5.4 Chapter summary

With the growing adoption of IIoT and advancements in robot learning, predictable intelligence in HRC holds the potential to greatly enhance flexible production capabilities for intricate manufacturing tasks in the HCSM environment. As a foundational step, the prediction of ongoing human intentions should leverage short-time frames while incorporating multiple modalities of information. Then, through the transfer of knowledge from daily activities to assembly patterns, online operation intention analysis can be conducted. Based on these predictions, dynamic decisions can be made in advance, enabling a mobile robot to assist human operators in a proactive and predictable manner.

In summary, this chapter makes scientific contributions in two key areas:

a) It introduces a multimodal fusion network designed to predict ongoing human intentions with a reduced number of incoming frames. This approach addresses a challenge in industrial action pattern recognition, particularly concerning diverse intricate, fine-grained operations involved.

b) It presents a framework that enables predictable intelligence in Proactive HRC through online intention prediction and proactive robot motion. This approach empowers a robot to anticipate human intentions in advance, facilitating collaborative work that aligns seamlessly with human expectations.

References

Baradel, F., Wolf, C., Mille, J., Taylor, G.W., 2018. Glimpse clouds: Human activity recognition from unstructured feature points. In: Proceedings of the IEEE Conference on Computer Vision and Pattern Recognition, pp. 469–478.

Cao, Z., Simon, T., Wei, S.E., Sheikh, Y., 2017. Realtime multi-person 2D pose estimation using part affinity fields. In: Proceedings of the IEEE Conference on Computer Vision and Pattern Recognition, pp. 7291–7299.

Carreira, J., Zisserman, A., 2017. Quo vadis, action recognition? A new model and the kinetics dataset. In: Proceedings of the IEEE Conference on Computer Vision and Pattern Recognition, pp. 6299–6308.

Dang, Y., Yang, F., Yin, J., 2020. DWnet: Deep-wide network for 3D action recognition. Robotics and Autonomous Systems 126, 103441.

De Boissiere, A.M., Noumeir, R., 2020. Infrared and 3D skeleton feature fusion for RGB-D action recognition. IEEE Access 8, 168297–168308.

Fan, J., Li, S., Zheng, P., Lee, C.K., 2021. A high-resolution network-based approach for 6D pose estimation of industrial parts. In: 2021 IEEE 17th International Conference on Automation Science and Engineering (CASE). IEEE, pp. 1452–1457.

Li, S., Fan, J., Zheng, P., Wang, L., 2021a. Transfer learning-enabled action recognition for human–robot collaborative assembly. Procedia CIRP 104, 1795–1800.

Li, S., Zheng, P., Fan, J., Wang, L., 2021b. Toward proactive human–robot collaborative assembly: A multimodal transfer-learning-enabled action prediction approach. IEEE Transactions on Industrial Electronics 69, 8579–8588.

Liu, G., Qian, J., Wen, F., Zhu, X., Ying, R., Liu, P., 2019. Action recognition based on 3D skeleton and RGB frame fusion. In: 2019 IEEE/RSJ International Conference on Intelligent Robots and Systems (IROS). IEEE, pp. 258–264.

Liu, H., Wang, L., 2021. Collision-free human–robot collaboration based on context awareness. Robotics and Computer-Integrated Manufacturing 67, 101997.

Shahroudy, A., Liu, J., Ng, T.T., Wang, G., 2016. NTU RGB+D: A large scale dataset for 3D human activity analysis. In: Proceedings of the IEEE Conference on Computer Vision and Pattern Recognition, pp. 1010–1019.

Su, L., Hu, C., Li, G., Cao, D., 2020. MSAF: Multimodal split attention fusion. arXiv: 2012.07175.

Wang, B., Hu, S.J., Sun, L., Freiheit, T., 2020. Intelligent welding system technologies: State-of-the-art review and perspectives. Journal of Manufacturing Systems 56, 373–391.

Wen, X., Chen, H., Hong, Q., 2019. Human assembly task recognition in human–robot collaboration based on 3D CNN. In: 2019 IEEE 9th Annual International Conference on CYBER Technology in Automation, Control, and Intelligent Systems (CYBER). IEEE, pp. 1230–1234.

Yan, S., Xiong, Y., Lin, D., 2018. Spatial temporal graph convolutional networks for skeleton-based action recognition. In: Proceedings of the AAAI Conference on Artificial Intelligence.

Zheng, P., Li, S., Xia, L., Wang, L., Nassehi, A., 2022. A visual reasoning-based approach for mutual-cognitive human–robot collaboration. CIRP Annals.

CHAPTER 6

Self-organizing multi-agent teamwork

Self-organizing task allocation is vital for collaboration among multiple human and robotic agents to satisfy changing operational objectives and workspace dynamics. Nonetheless, current HRC systems heavily rely on predefined task configurations for human and robot agents, fail to consider manufacturing requirements from diverse operation sequences and varying mechanical components. To tackle this challenge, this chapter introduces a temporal subgraph approach for the task planning in Proactive HRC settings with multiple agents. The task allocation strategy is represented using a tri-layer KG that captures the relationships among tasks, agents, and operations. Simultaneously, we incorporate a temporal subgraph reasoning mechanism to extract implicit and historical knowledge from the KG, anticipating forthcoming actions for both humans and robots. To showcase the efficacy of the proposed methodology, we applied it to an assembly tasks of car engine and gearbox, respectively. The experimental results achieved significant performance.

6.1 Connotation

In the pursuit of HCSM paradigm (Lu et al., 2022), research explorations have been ignited to enhance human well-being and foster adaptable production processes (Kovalenko et al., 2021). A pivotal element in achieving hybrid collaboration between humans and robots in manufacturing tasks is task allocation. For instance, Liau and Ryu (2022) decomposed assembly operations into substage actions, studying their agent assignment preferences. A genetic algorithm was employed for task allocation in HRC systems with two robots. Similarly, Karami et al. (2020) utilized an AND–OR graph to delineate task attributes, operator capabilities, and robot functionalities, facilitating the sequential allocation of tasks among a human operator, a mobile manipulator, and a dual-arm manipulator for defect inspection tasks.

Nonetheless, it is noted that the task decomposition approach is only feasible for the generation of fixed sequential operations, lacking adaptabil-

Proactive Human–Robot Collaboration Toward Human-Centric Smart Manufacturing
https://doi.org/10.1016/B978-0-44-313943-7.00013-2

ity during multi-agent collaboration. The collaborative processes among multiple human and robot agents fail to achieve self-organize dynamic task assignments and resource allocation when confronted with changes in the workspace. Moreover, existing HRC settings predominantly involve a solitary human operator and a single robot, which cannot address intricate tasks that rely on the collaborative efforts of multiple human–robot agents. For example, tasks such as assembling large-scale complex products often require the coordination of human operators assuming diverse roles and robots with varying payloads (Liu et al., 2020).

Motivated by these challenges, there is a growing interest in the self-organizing completion of collaborative tasks involving multiple humans and robots. Self-organizing intelligence in HRC empowers dynamic task planning, enabling multiple humans and robots to cooperatively execute manufacturing activities while considering their unique capabilities and task policies. Understanding task structures and interpreting changes in the workspace offer latent cues for fostering self-organizing capabilities in multi-agent HRC scenarios.

Figure 6.1 The connotation of self-organizing intelligence in Proactive HRC.

In the implementation of such HRC systems, several challenges necessitate efficient solutions, as illustrated in Fig. 6.1. Firstly, two issues are in the real-time perception of on-site HRC settings, including: 1) the acquisition of real-time status concerning humans, robots, and components, and 2) the detection of workspace changes during various task stages. The utilization of a graph R-CNN algorithm (Yang et al., 2018) can serve to detect various objects and establish their relationships within HRC scenes. Additionally, a subgraph reasoning mechanism within the KG can focus on changes in geometric information of component sets.

The subsequent module revolves around task knowledge representation and focuses on two aspects: 1) the identification of relationships between perceived objects and HRC tasks, and 2) the dynamic adjustment of task planning strategies across different timestamps. In this context, a KG spanning task–agent–operation layers can be introduced to establish connections between human, robot, component, and action nodes, extracting both explicit and implicit knowledge regarding HRC tasks and the diverse capabilities of multiple agents. Furthermore, a temporal reasoning mechanism can be incorporated to learn from prior knowledge and adapt it to evolving task stages, thereby updating task planning strategies.

Lastly, exploring KGs with graphical structures can offer an explanatory means for task allocation involving multiple agents. This graphical representation simplifies the depiction of sequential task sequences in an easily understandable and intuitive manner.

6.2 A temporal subgraph neural network-based framework

This chapter introduces an architecture designed to facilitate self-organizing intelligence within multiple HRC. Specifically, it employs a temporal subgraph approach within a KG to develop self-organizing task planning strategies by understanding task structure and geometric interpretations of the HRC workspace. The HRC KG also serves as a graphical representation for elucidating operation sequences across various task stages. Within this framework, an action node denotes discrete robot motion types, while a component node depicts the endpoint of motion trajectories. Hence, robots can autonomously execute manipulations guided by KG instructions. Simultaneously, human operators gain insight into changing object relationships within the workspace and receive reasonable operation recommendations through the continuously updated KG.

The overall architecture of this self-organizing intelligence system, involving multiple human–robot agents, is represented in Fig. 6.2. Real-time visual data from HRC workspaces is obtained using a depth camera to monitor task execution progress over time. Leveraging this visual input, a graph R–CNN algorithm identifies and tracks human(s), robots, and components, representing them as nodes (entities) within the KG. This dynamic node interaction within different timestamps contributes to the construction of a temporal SG, an integral component of the HRC KG. The HRC KG encompasses human, robot, component, and action nodes, with historical nodes from previous HRC execution stages retained within the graph.

On the one hand, the subgraph reasoning module integrates various subregions of node sets within the KG to infer HRC task planning knowledge across task–agent–operation levels. On the other hand, the temporal

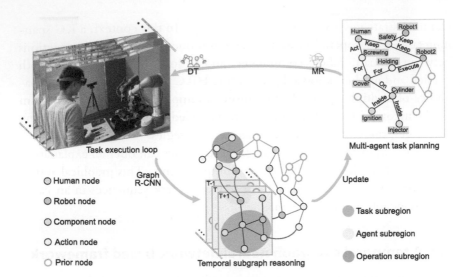

Figure 6.2 The framework of self-organizing HRC between multiple humans and robots. (Adapted from (Li et al., 2023).)

reasoning module extracts temporal insights from different task stages by tracing back through historical KG data and prior nodes, enabling updates to the HRC KG. Subsequently, multi-agent task planning is generated based on the updated HRC KG, facilitating inference of both human and robotic operations. This enables fluent HRC task execution and efficient collaboration. The generated task planning strategies are communicated to human and robotic agents in MR environments.

6.2.1 Tri-layer HRC KG for time-changing task allocation

To enable dynamic task allocation in self-organizing HRC involving multiple agents, a temporal HRC KG is established, as depicted in Fig. 6.3. This KG interconnects nodes with diverse attributes through both intra-layer and inter-layer relations, forming a comprehensive representation of the HRC tasks. Throughout the task execution, the HRC KG retains historical operation records and continually updates the status of the human–robot–environment system, facilitating adaptive task assignment.

The task layer within the KG delineates relationships among various components and contains semantic knowledge of task fulfillment. The output of the graph R-CNN module, known as SG, embodies the structural framework of the task layer and remains interconnected with other subgraphs within the HRC KG. Across different timestamps, the task layer

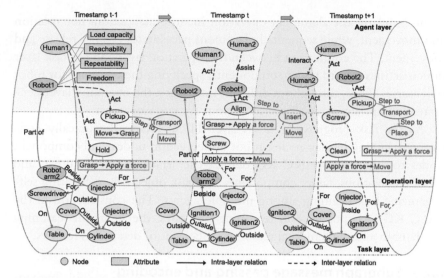

Figure 6.3 Temporal trilayer HRC KG for multi-agent task allocation in HRC. (Adapted from (Li et al., 2023).)

captures evolving relationships among component pairs, reflecting diverse stages of HRC tasks.

The agent layer characterizes the capabilities and roles of various agents within HRC teams composed of humans and robots. The priority for task allocation to a human operator is assessed using three metrics: load capacity, reachability, and repeatability. These metrics, alongside the freedom degrees, define a robot's suitability for task execution. Over time, the agent layer computes varying reachability levels for both humans and robots based on their changing proximity to target components. Additionally, the load capacity and repeatability of human operations are adjusted dynamically as human operators engage in different task stages. Active and supportive roles are flexibly assigned to agents as needed during different stages in collaborative operations.

The operation layer dissects potential human–robot operations into sequences of atomic actions and links these actions to various task executions. These atomic actions can be assigned to and executed by both human and robotic agents. For instance, in assembling a fuel injection tube into an engine cylinder head, a robot's tasks include picking up the tube, transporting it over the cylinder head, while a human operator assists in alignment and placement. The picking operation involves actions like moving and grasping, while the transport procedure entails applying force and movement. Similarly, the alignment operation incorporates actions like rotation and movement, while sequential placement necessitates gripper release and movement. The connection between the operation layer and the task layer directs actions toward target components, specifying the endpoint for robot trajectories. Thus, throughout different task stages, the operation layer generates appropriate sequences of operations for task execution.

As the task progresses, the tri-layer HRC KG is dynamically constructed, linking human nodes, robot nodes, action nodes, and component nodes together. This HRC KG defines the arrangement of human and robotic operations, target components, and task objectives, facilitating self-organizing task planning among multiple agents. The construction of the HRC KG incorporates two reasoning processes: the subgraph reasoning mechanism and the temporal reasoning mechanism, detailed as follows.

6.2.2 Subgraph message passing and encoding

In the context of temporal HRC KG, denoted as $G = (V, E)$ at timestamp t, the node set V encompasses humans, robots, components, and action entities, with their relations described by the edge set E. A subgraph, referred to as $S = (V', E')$, is employed to represent nodes and their relationships across task, operation, and agent layers. The subgraph encoding technique is utilized to generate node embedding representations, facilitating link prediction by capturing graph topology among various subgraphs, denoted as $S = (S_1, S_2, \ldots, S_n)$. This subgraph representation emerges from three successive steps: initial message propagation from a node v_i to its neighbor $v_j \in \mathcal{N}_{v_i}$, aggregation of these messages from all neighbors with different weights, and encoding of the aggregated message using an activation function, thereby learning hidden feature representations for layer-wise updates.

The subgraph's topology includes both internal and border connections at position, neighbor, and structure levels, as illustrated in the left section of Fig. 6.4. The internal position denotes the distances between nodes within the same subgraph, whereas the border position quantifies the distances reaching nodes beyond the subgraph. The internal neighbor identifies nodes directly linked within a subgraph, whereas the border neighbor includes nodes from the entire KG with the highest relatedness, ranking the top k scores. The internal structure represents the connectivity of node sets within each subgraph, while the border structure connects a node within a subgraph to its neighbors beyond the subgraph. In this manner, information from the temporal HRC KG is captured from multiple distinct subgraphs.

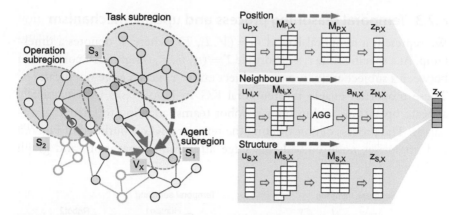

Figure 6.4 The mechanism of subgraph message passing and encoding. (Adapted from (Li et al., 2023).)

For subgraph S, the message channels originating from the position, neighbor, and structure attributes to a node V_X are denoted as $U_{P,X}$, $U_{N,X}$, and $U_{S,X}$, respectively, as illustrated in the left section of Fig. 6.4. In essence, the message propagated to node V_X from each subgraph S_i is computed as follows:

$$MSG_i^{S \to V_X} = W_i \cdot (V_X, S_i). \tag{6.1}$$

Here, we introduce learnable weight parameters, denoted as W_i, which are associated with different subgraphs. The message-passing procedure yields representations, namely $M_{P_i,X}$, $M_{N_i,X}$, and $M_{S_i,X}$, for the three subgraph channels. Subsequently, these messages are aggregated across all subgraphs. In the l-th layer, a nonlinear activation function, represented by σ, is employed to transform the aggregated messages from the position, neighbor, and structure channels into a hidden representation for the subsequent layer iteration. This process can be presented as follows:

$$a_{i,X} = AGG(MSG_i^{S \to V_X} \ \forall S_i \in S),$$
$$Z_{i,X}^{l+1} \leftarrow \sigma(a_{i,X}; Z_{i,X}^l). \tag{6.2}$$

Through layer-wise message passing, we encode the internal and border characteristics of subgraphs from three channels, yielding property-aware output representations. These representations are denoted as $\{Z_{P_I,X}, Z_{P_B,X}\}$, $\{Z_{N_I,X}, Z_{N_B,X}\}$, and $\{Z_{S_I,X}, Z_{S_B,X}\}$. Ultimately, these representations are combined through concatenation to generate the node embedding Z_X, including both positional and structural information pertaining to subgraphs.

6.2.3 Temporal reasoning process and update mechanism

We represent a temporal KG as $G = (V, E, T)$, where T denotes a timestamp. At a specific time t, a quadruple $\mathcal{I} = (v_s, p_j, v_o, t)$ signifies a connection between a subject entity v_s and an object entity v_o via an edge (predicate) p_j. These properties enable the temporal KG in HRC tasks to dynamically allocate operations within human–robot teams, adapting to changing scenarios. The process of constructing the temporal KG is outlined in Fig. 6.5 and comprises node embedding, edge attention propagation, and graph updating.

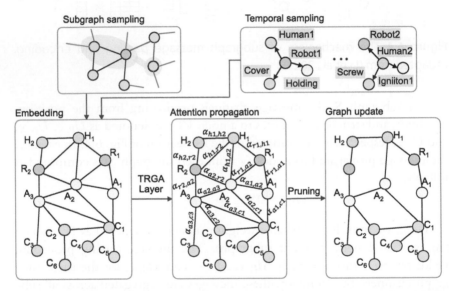

Figure 6.5 The mechanism of temporal graph reasoning and update. (Adapted from (Li et al., 2023).)

The node embedding module integrates information from both temporal sampling and subgraph sampling. Quadruples denoting past human–robot–component–action relations are used to construct temporal node embeddings, as illustrated in the upper left corner of Fig. 6.5. For a node v_s, its prior facts $\mathcal{Q} = (v_s, p_k, v_p, t' | t' < t)$ are collected to share knowledge of accomplished HRC operations across timestamps. The temporal embedding of node v_s is computed through message passing from its set of prior neighbors $\hat{N}_{v_s}$, as denoted in Eqs. (6.1) and (6.2). Meanwhile, the node subgraph embedding is derived from the previously described subgraph encoding process. The concatenation of the temporal embedding and subgraph embedding results in evolving entity features that exhibit subgraph patterns. This time-aware and subgraph–based node embedding is expressed as $Z_s(x, t) = [\bar{Z}_x || Z_t]^T \in \mathbb{R}^{d_X + d_T}$. Here, $\bar{Z}x \in \mathbb{R}^{d_X}$ represents the subgraph embedding that allows an entity to access subregion information within the HRC KG, where d_X signifies the feature dimensionality. Simultaneously, $Zt \in \mathbb{R}^{d_T}$ captures time-evolving information from prior facts Q and learns the progress of HRC tasks by considering the temporal graph structure. Lastly, an embedding vector p_k is learned from the updated graph structure to represent predicate features between different entities.

Following the embedding process, the Temporal SubGraph Attention (TSGA) layer is introduced to distill a graph representation that incorporates temporal and subgraph relations. This layer takes entity embedding and predicate embedding as input and generates an edge attention score from a subject entity to an object entity, resulting in new hidden representations for node sets. The edge attention score assigns varying levels of importance to different edges, calculated as follows:

$$e_{so}^l(z_{so}, p_j) = W_s^l(h_s^{l-1} || p_j^{l-1}) W_o^l(h_o^{l-1} || p_j^{l-1}). \tag{6.3}$$

In this context, e_{so}^l represents the edge attention score pertaining to the quadruple $\mathcal{I}$. The node pair z_{so} encompasses both z_s and z_o, while h_s^{l-1} and h_o^{l-1} denote the hidden representations at layer $l-1$ corresponding to the node embeddings $Z_s(x, t)$ and $Z_o(x, t)$, respectively. We also have the predicate embedding p_j and two learnable weight matrices W_s and W_o. The edge attention mechanism is used to blend predicate features with entity features, capturing relational information within the graph. Subsequently, the attention score is normalized through the following process:

$$\alpha_{so}^l(z_{so}, p_j) = \frac{\exp\left(e_{so}^l(z_{so}, p_j)\right)}{\sum_{v_c \in \mathcal{N}_{v_s}} \sum_{p_i \in \mathcal{P}_{sc}} \exp\left(e_{sc}^l(v_s, v_c)\right)}. \tag{6.4}$$

Here, we use $\mathcal{N}_{v_s}$ to represent the neighbors of entity v_s, and $\mathcal{P}_{sc}$ refers to the edge set connecting node v_s and node v_c. The subsequent stage involves aggregating the hidden representations of these neighbors, each with its own weight score, as indicated by the following equation:

$$\hat{h}_s^l(z_s) = \sum_{v_c \in \mathcal{N}_{v_s}} \sum_{p_i \in \mathcal{P}_{sc}} \alpha_{sc}^l(z_{sc}, p_i) h_c^{l-1}(v_c). \tag{6.5}$$

Next, we apply the LeakyReLU activation function to refresh the node embedding of z_s by activating both its layer-wise hidden representation and the combined features from its neighbors. Concurrently, we update the predicate embedding to prepare for subsequent layer-wise representation aggregation. This update process within the TSGA layer is represented as follows:

$$
\begin{aligned}
h_s^l(z_s) &= \sigma(W_h^l(\beta h_s^{l-1}(z_s) + (1-\beta)\hat{h}_s^l(z_s) + b_n^l)), \\
p_i^l &= W_h^l p_i^{l-1} + b_h^l, \quad \forall p_i \in \mathcal{P}_{sc}.
\end{aligned}
\tag{6.6}
$$

In the equation above, we introduce β as a hyperparameter that modulates the impact of hidden and aggregated features during embedding updates. Additionally, we have W_h^l as the weight matrix and b_h^l as the bias vector within the activation function.

After the HRC KG undergoes the TSGA layer, different attention scores are assigned to various graph edges. These propagated edge attention scores are used to compute the attention value for node v_s at the l layer, as expressed below:

$$
c_{v_s}^l = \sum_{v_c \in \mathcal{N}_{v_s}} \sum_{p_i \in \mathcal{P}_{sc}} \alpha_{sc}^l(z_{sc}, p_i) c_{v_c}^{l-1}.
\tag{6.7}
$$

Next, we combine the edge attention score and the node attention value to determine the contribution of the edge associated with predicate p_j between entities v_s and v_o. This contribution is represented as $a_{so}(p_j) = \alpha_{so}^l(z_{so}, p_j) c_{v_s}^l$. Subsequently, we perform a pruning step by retaining the top K edges. These edges are selected based on the ranking of their contribution scores. The HRC KG is then updated and reconstructed using these selected edges that connect different nodes.

For the supervision training process, we employ binary cross-entropy as the loss function to facilitate the learning of node embeddings and optimize model parameters during the message-passing iterations. This loss function is defined as follows:

$$
\begin{aligned}
L = -\frac{1}{|\mathcal{P}|} \sum_{p_i \in \mathcal{P}_{si}} \frac{1}{|\mathcal{P}_{si}|} \sum_{v_i \in \mathcal{N}_{v_s}} & (y_{v_i, p_i} \log(\frac{c_{v_i}}{\sum_{v_j \in \mathcal{N}_{v_s}} c_{v_j}}) \\
& + (1 - y_{v_i, p_i}) \log(1 - \frac{c_{v_i}}{\sum_{v_j \in \mathcal{N}_{v_s}} c_{v_j}})).
\end{aligned}
\tag{6.8}
$$

In this context, we utilize $\mathcal{P}$ to represent the set of quadruples within the graph. Meanwhile, $\mathcal{P}_{si}$ stands for the set of node pairs comprising entity v_s and its neighbors $v_i \in \mathcal{N}_{v_s}$, with p_i serving as the predicate connecting these node pairs. Additionally, c_{v_i} represents the node attention value assigned to entity v_i, and y_{v_i,p_i} is the label that indicates whether v_i is the object entity connected to v_s through the predicate p_i.

6.3 Case study

In this section, we present a case study focused on self-organizing task allocation among multiple human and robotic agents, with the objective of collaboratively assembling a partial car engine and gearbox, respectively. To assess the effectiveness of our proposed temporal subgraph reasoning method, we conduct comparative experiments, evaluating the performance of our approach in HRC context.

6.3.1 Temporal subgraph explanation for HRC assembly

HRC plays a vital role in enhancing the working conditions during the assembly of car engines. In this assembly process, robots are employed to handle high-payload and repetitive tasks, which significantly reduces the risk of musculoskeletal disorders for human operators. Simultaneously, human operators are crucial for performing delicate operations such as screwing and wire organization. The HRC workspace for car engine assembly is depicted in Fig. 6.6. Here, the system contains an ABB robot, a UR robot, and a human operator working closely alongside partial engine parts and various tools. The human operator receives real-time information and procedural guidance through HoloLens2, while robot controllers receive instructions via Ethernet. The assembly process involves various components and tools, including a valve cover, cylinder head, four ignition coils, a fuel injector tube, numerous screws, a screwdriver, and a hex key. An Azure Kinect camera captures RGB-D visual data at 30 Hz, providing input for scenario analysis and task planning.

The assembly process can be divided into three subtasks. These temporal subgraphs help elucidate the operational sequences within these subtasks. Each action entity linked to a robot entity represents robotic motion types, such as moving or grasping. Subsequently, the connection from the action node to a component node indicates the endpoint of the robot's trajectory. This approach allows robots to plan and execute continuous manipulations according to the graphical structure. Similarly, the connection from a human entity to an action entity, followed by component nodes, provides operation instructions for human tasks. The node pairs between

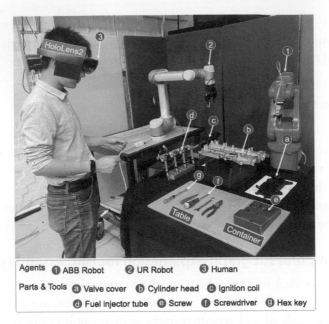

Figure 6.6 HRC workspace settings for assembly tasks of car engines. (Adapted from (Li et al., 2023).)

different component entities describe their spatial relationships within the workspace.

The first subtask focuses on inserting the fuel injector tube into the cylinder head and fastening screws for these components, illustrated in Fig. 6.7. A temporal subgraph spanning four timestamps is employed to represent the assembly operation sequences, which can be described as follows:

i) A robot node is linked to the *picking* entity, comprising moving and grasping atomic actions, followed by pointing to the *fuel injector tube* component entity. A human node is connected to the *taking* entity, sequentially pointing to the *screw* and the *screwdriver* entities.

ii) A robot node links to the *transporting* entity, involving actions like applying force and moving, followed by pointing to the *cylinder head* entity.

iii) A robot node is associated with the *aligning* entity, which includes rotating and moving atomic actions, followed by pointing to the *cylinder head* entity. This robot manipulation aligns the four pipe pins of the fuel injector tube with the holes in the cylinder head.

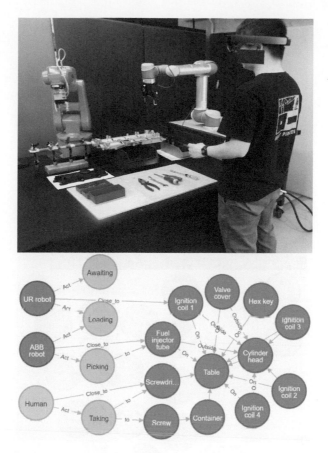

Figure 6.7 Temporal subgraph representation for assembly sequences in the first stage. (Adapted from (Li et al., 2023).)

iv) A robot node is connected to the *placing* entity, involving actions like releasing the gripper and moving backward, followed by pointing to the *fuel injector tube* entity. A human node is engaged in the *screwing* entity and subsequently points to the *fuel injector tube* entity.

The second subtask entails inserting four ignition coils into the cylinder head, followed by screw mounting operations, as shown in Fig. 6.8. The assembly sequences for this stage include:

i) A robot picks up an ignition coil, while another robot ensures collision avoidance with the fuel injector tube, and the human takes screws, switching tools to the hex key if necessary.

ii) The robot transports the ignition coil to the cylinder head.

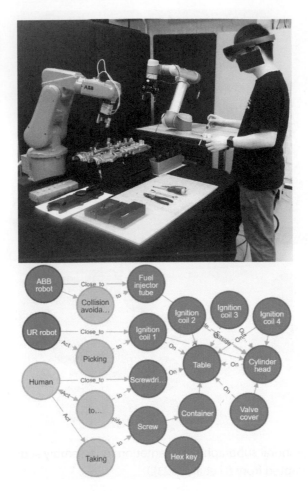

Figure 6.8 Temporal subgraph representation for assembly sequences in the second stage. (Adapted from (Li et al., 2023).)

iii) The robot places the ignition coil into a hole in the cylinder head, while the human secures it in place with screws. The process is encapsulated within a temporal subgraph spanning three timestamps.

The *collision avoidance* entity in the first timestamp guides the robot's backward movement to maintain a safe distance from the target object, ensuring the mounting of all four ignition coils onto the cylinder head.

The final subtask involves assembling the valve cover onto the cylinder head (see Fig. 6.9), with the following steps:

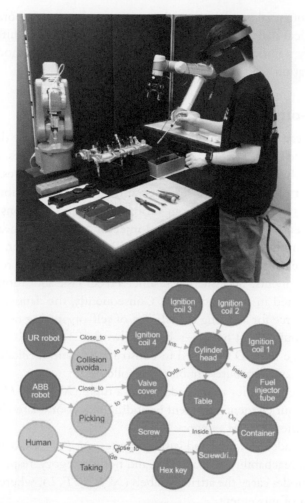

Figure 6.9 Temporal subgraph representation for assembly sequences in the third stage. (Adapted from (Li et al., 2023).)

i)   Two robots pick up the valve cover, ensuring collision avoidance with the ignition coils, while the human operator retrieves screws.
ii) One robot transports the valve cover above the cylinder head.
iii) The robot performs the holding action, while the human guides the robot to align the bolt holes between the valve cover and the cylinder head.
iv) The robot places the valve cover, while the human secures it in place with screws.

These sequential operations are also represented using temporal subgraph-based graphical structures spanning four timestamps. In the third step, the *holding* entity signifies that the robot switches to compliance mode to facilitate cooperation with the human operator(s).

6.3.2 Self-organizing task allocation result

To facilitate car engine assembly across its three stages, we have developed a Task Sequence Dataset (TSD) comprising 2202 RGB-D images. Here 1531 images with their associated labels are used for training purposes, while the remainder serves as the test set. Each specific timestamp and subtask within TSD captures multiple images from various angles and locations relative to the components. For instance, the fuel injector tube may appear near the ABB robot in one image and next to the UR robot in another. These images, within each timestamp, present four different geometric relationships among components, necessitating distinct HRC task allocation strategies that are reflected in the annotations. Consequently, the dataset serves as a valuable resource for assessing the efficacy of self-organizing task allocation methods in multi-agent human–robot teamwork.

Each image in TSD is labeled to construct KG and facilitate task allocation, with node attributes annotated as follows:

- Component entities are recorded with a tuple (X, Y, W, H, L), where the first four elements represent bounding box coordinates, and the last one signifies the part category.
- Human nodes are annotated with attributes (N, P, R, L), denoting load capacity, reachability, repeatability, and node category, respectively.
- Robot nodes carry the attribute labels (N, P, R, F, L), where F denotes freedom of movement.

Specifically, maximum payloads for humans, ABB robots, and UR robots are set at 2, 4, and 5 kg, respectively. Human reachability is defined as the minimum distance from human hands to target objects, while robot reachability is measured as the distance from the robot gripper to the manipulated component. Repeatability for both human and robot agents is set to a constant value of 1, considering that the assembly task can be completed in ten steps. The freedom of movement for ABB and UR robots is 6 DoF.

In training the temporal subgraph model, the agent subregion comprises human and robot nodes, while the sets of component nodes and action nodes represent the task and operation subregions, respectively. We employ three hidden representation layers, and the model is trained with a learning rate of 0.0002 on a Tesla V100 GPU (16 GB) with a batch size of 32, utilizing SGD as the optimizer. For the TSGA layer update, we set the hyperparameter β to 0.5 and the parameter K to 3. Additionally, edges with contribution scores exceeding the 0.6 threshold value are retained during the pruning stage.

During the testing phase, the SG constructed by the Graph R–CNN model serves as the prior knowledge for the first timestamp of each assembly subtask. Leveraging this observed information, the trained temporal subgraph model is applied to complete the entire KG for HRC task allocation. It forecasts the correct action nodes linked to human and robot nodes, along with the subsequent target component nodes. Subsequent timestamps within an assembly subtask update the HRC KG using the trained model to learn node embeddings from the timely observed SG and the prior knowledge.

To evaluate our proposed temporal subgraph model's performance, we employ standard metrics, including Mean Reciprocal Ranking (MRR), Hits@1, and Hits@3 (Han et al., 2021), comparing our approach with other benchmarks. The evaluation encompasses triples from agents to action nodes and from actions to component nodes in all testing images. The diverse geometric configurations and evolving relationships between components in the test data validate the effectiveness of the self-organizing task allocation model, which forecasts human–robot sequential actions.

Across the three subtasks of car engine assembly, the comparison results between our model and GCN, GAT (Veličković et al., 2017), and RGAT (Busbridge et al., 2019) are presented in Tables 6.1, 6.2, and 6.3. The quantitative findings demonstrate that our model significantly enhances link prediction accuracy for humans, robots, and component nodes, underscoring its ability to extract implicit information from subregions and temporal procedures. Moreover, in all three assembly subtasks, accuracy in subsequent timestamps surpasses that of the first timestamp, indicating that our model benefits from observing more prior facts as the task progresses.

To further evaluate the effectiveness of our method, an ablation study is conducted (Table 6.4). Reducing the graph pruning threshold negatively impacts model performance since most nodes in the graph have multiple edges. Removing either the subgraph sampling or temporal sampling process also leads to a decline in link prediction accuracy, verifying the positive contributions of these mechanisms.

Table 6.1 Results of link prediction on the first subtask of car engine assembly. Evaluation metrics are MRR (%) and Hits@1/3 (%).

Subtask1	Timestamp 1			Timestamp 2			Timestamp 3			Timestamp 4		
Model	MRR	Hits@1	Hits@3	MRR	Hits@1	Hits@3	MRR	Hits@1	Hits@3	MRR	Hits@1	Hits@3
GCN	89.10	82.51	90.43	87.79	81.96	86.89	88.01	81.15	87.70	88.99	82.79	89.75
GAT	92.46	86.89	95.08	91.47	86.88	91.81	91.15	86.89	89.34	91.41	86.48	92.21
RGAT	93.47	88.52	96.45	92.09	87.70	92.62	91.01	86.89	89.34	91.94	87.30	92.62
Ours	94.67	90.44	97.27	95.28	92.63	95.08	95.04	92.62	95.08	95.55	92.62	96.72

Table 6.2 Results of link prediction on the second subtask of car engine assembly. Evaluation metrics are MRR (%) and Hits@1/3 (%).

Subtask2	Timestamp 1			Timestamp 2			Timestamp 3		
Model	MRR	Hits@1	Hits@3	MRR	Hits@1	Hits@3	MRR	Hits@1	Hits@3
GCN	88.34	82.17	87.91	88.18	80.33	90.16	88.51	81.97	88.52
GAT	90.95	86.27	90.57	90.57	84.42	91.80	91.04	86.07	91.39
RGAT	92.61	88.73	92.62	92.90	88.52	93.44	91.96	87.30	92.62
Ours	94.39	90.98	94.88	95.70	92.62	96.72	94.84	91.80	95.49

Table 6.3 Results of link prediction on the third subtask of car engine assembly. Evaluation metrics are MRR (%) and Hits@1/3 (%).

Subtask3	Timestamp 1			Timestamp 2			Timestamp 3			Timestamp 4		
Model	MRR	Hits@1	Hits@3	MRR	Hits@1	Hits@3	MRR	Hits@1	Hits@3	MRR	Hits@1	Hits@3
GCN	88.92	82.51	88.52	87.99	80.33	88.52	90.05	83.61	90.98	88.36	80.73	90.16
GAT	91.18	85.79	91.80	90.71	84.42	91.80	92.00	86.89	92.62	91.84	86.48	93.85
RGAT	92.21	87.70	92.08	92.26	87.70	93.44	93.23	88.93	93.44	93.06	88.11	95.49
Ours	93.88	90.16	94.54	94.88	91.80	94.26	95.70	92.65	96.73	95.90	93.03	97.13

Table 6.4 Ablation study results on three subtasks. Evaluation metric is MRR (%).

Model	Subtask 1	Subtask 2	Subtask 3
Proposed ($K = 3$)	95.06	94.71	94.97
Proposed ($K = 5$)	93.35	92.94	93.72
Proposed ($K = 1$)	89.84	90.36	89.67
w/o subgraph sampling	92.13	92.86	92.45
w/o temporal sampling	91.68	92.38	91.94

6.3.3 Demonstration on a gearbox assembly task

The temporal subgraph methodology can also be applied in a gearbox assembly operation, via the collaborative efforts of a human operator, a UR5 robot, and a UR5e robot. This HRC scenario is illustrated in Fig. 6.10. The assembly task comprises 12 mechanical components, including the bottom cover, pump body, driven gear, driving gear, up cover, sealing ring, groove, front shelf, plate shelf, right shelf, and conveyor, all arranged on the shared workbench. The overall assembly process is compartmentalized into eight discrete subtasks, with each subtask introducing changes in the spatial configuration and interrelation of these components. To effectively manage these evolving dynamics, the temporal subgraph consistently preserves historical data records and anticipates operation scheduling within the HRC KG.

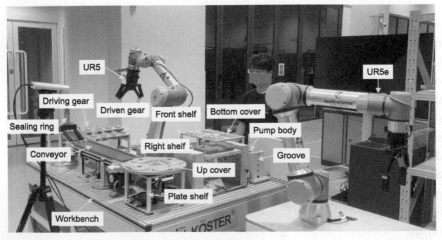

Figure 6.10 The workspace setting of gearbox assembly task with multiple HRC.

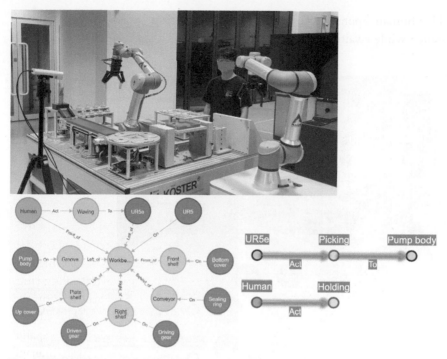

Figure 6.11 The gearbox assembly task with multiple HRC operation sequences – step 1.

As illustrated in Fig. 6.11, the first step depicts the HRC settings for the assembly process. Here, the human operator takes a position at the forefront of the workbench, engaging in a waving gesture directed toward the UR5e robot, which is located to the left of the workbench. Simultaneously, the UR5 robot is affixed to the workbench. Various components of the gearbox are arranged as follows: the bottom cover is placed on the front shelf, fastened to the front edge of the workbench; and the sealing ring is situated on the conveyor, located at the rear of the workbench. Both the driving gear and the driven gear are located on the right shelf, anchored to the right-hand side of the workbench. Furthermore, the up cover of the gearbox resides on the plate shelf, which is fixed to the left of the workbench, and the pump body is positioned on the groove, which is located on the left-hand side of the workbench. Within this spatial arrangement, the temporal subgraph model forecasts that the UR5e robot is tasked with picking the gearbox's pump body and conveying it to the human operator.

The human operator prepares for this interaction by adopting a holding stance while awaiting the robot's delivery.

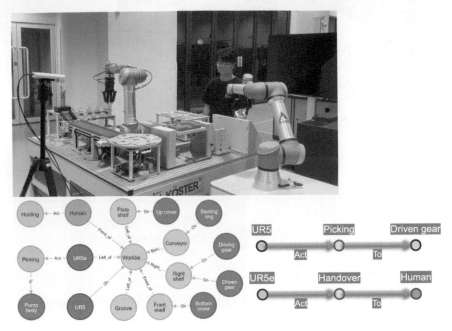

Figure 6.12 The gearbox assembly task with multiple HRC operation sequences – step 2.

Then, as elucidated in step 2, Fig. 6.12 shows a shift in relationships of HRC elements. During this phase, the human operator assumes a holding posture while the UR5e robot performs the task of retrieving the gearbox's pump body. Following these coordinated actions, the reasoning mechanism deduces that the UR5 robot should now engage in picking the driven gear from the right shelf. Meanwhile, the UR5e robot executes a handover action with the human operator, adapting the collaborative dynamics of the assembly process.

In the subsequent phase, denoted as step 3 in Fig. 6.13, the human operator undertakes the task of assembling the pump body with the bottom cover of the gearbox. Meanwhile, the UR5 robot executes the retrieval of the driven gear. In light of these observed interactions, the inference model dictates that the UR5 robot is now tasked with the precise placement of the driven gear onto the pump body. Simultaneously, the human operator

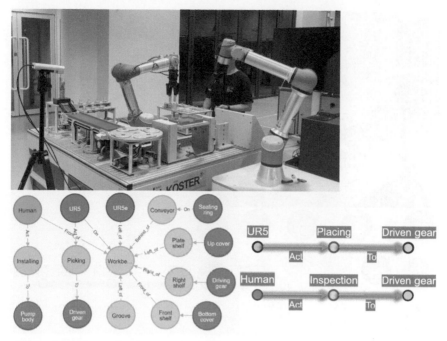

Figure 6.13 The gearbox assembly task with multiple HRC operation sequences – step 3.

assumes the role of inspector, ensuring the correct installation of the driven gear in the assembly process.

Then, as illustrated in Fig. 6.14 in step 4, the pump body is affixed to the bottom cover. Meanwhile, the driven gear is nestled within the pump body. In this scenario, the temporal subgraph model infers the forthcoming role of the UR5 robot, which should pick the driving gear for the progression of the assembly process.

With the assembly process progressed to step 5 in Fig. 6.15, the human–robot collaborative task execution continually changes. Within this stage, the human operator assumes the role of pointing the position of the up cover of the gearbox. Meanwhile, the UR5 robot undertakes the task of placing the driving gear within the confines of the pump body. The prediction results from reasoning model suggest that the UR5e robot should be assigned the responsibility of picking the up cover of the gearbox. In turn, the human operator should take on the role of inspecting the assem-

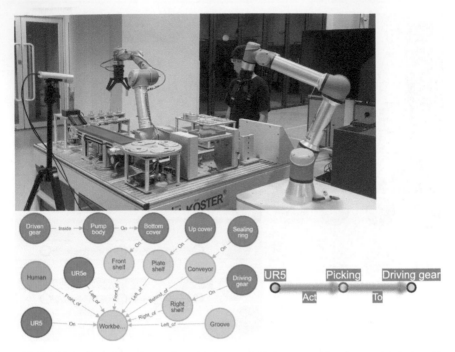

Figure 6.14 The gearbox assembly task with multiple HRC operation sequences – step 4.

bly process of the driving gear, thereby ensuring the quality and precision throughout the assembly process.

Advancing to the step 6 illustrated in Fig. 6.16, the UR5e robot grasps the up cover of the gearbox, paralleled by the holding stance of the human operator. Both the driven gear and the driving gear have been successfully nestled within the pump body. The graph model infers that the UR5e robot should participant the task of delivering the up cover to the human operator for a seamless handover operation.

Proceeding to step 7 depicted in Fig. 6.17, the human operator takes on the role of assembling the up cover to the pump body of the gearbox. Meanwhile, the UR5e robot smoothly navigates its way back to the left side of the workbench. In this particular subtask, the temporal subgraph model indicates that the UR5 robot's responsibility lies in picking the sealing ring from its position on the conveyor for the ongoing assembly process of the gearbox.

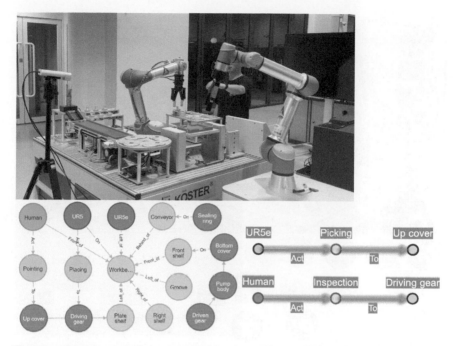

Figure 6.15 The gearbox assembly task with multiple HRC operation sequences – step 5.

In stage 8 depicted in Fig. 6.18, the UR5 robot takes hold of the sealing ring, while the human assumes a holding action, awaiting the next step. In the forthcoming operation, the reasoning model infers that the UR5 robot's task entails transferring the sealing ring to the human and executing the handover process. Following this, the human proceeds to install the sealing ring onto the gearbox, marking the successful completion of the gearbox assembly process.

6.4 Chapter summary

The concept of multiple HRC with self-organizing intelligence represents an emerging manufacturing paradigm. In this paradigm, multiple human and robotic agents collaborate to perform tasks that match their respective capabilities, enabling the accomplishment of complex operations. This chapter introduces a temporal subgraph method designed to infer relationships between humans, robots, actions, and components. This approach

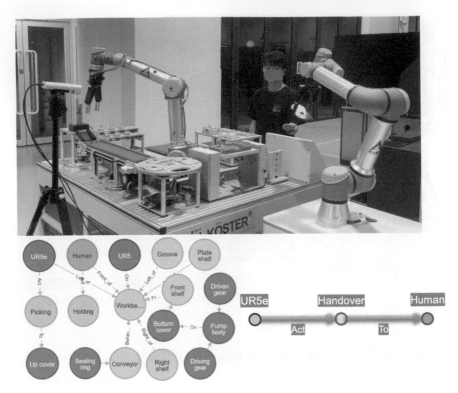

Figure 6.16 The gearbox assembly task with multiple HRC operation sequences – step 6.

facilitates the construction of an HRC KG that supports self-organizing task allocation. To summarize, this research contributes significantly in two key areas:

a) The temporal subgraph model can predict the next operations of human and robot agents. It achieves this by extracting knowledge representations from graph subregions and observed prior records. This capability allows for the dynamic adjustment of self-organizing task planning among multiple human and robot agents according to changes in task stages and the spatial locations of components.

b) The proposed tri-layer HRC KG serves as an explanatory framework for the arrangement of operations between humans and robots. Within this KG, robots can directly access information about motion types and destinations through intuitive graphical structures, while humans

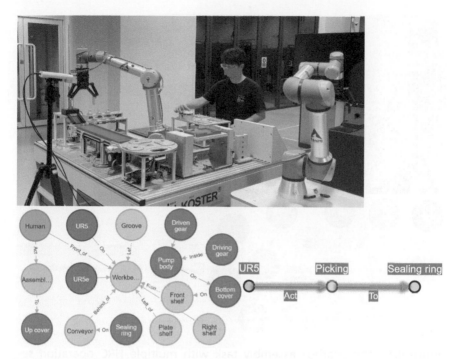

Figure 6.17 The gearbox assembly task with multiple HRC operation sequences – step 7.

gain insights into robot manipulation goals and receive suggestions for manual operations.

References

Busbridge, D., Sherburn, D., Cavallo, P., Hammerla, N.Y., 2019. Relational graph attention networks. arXiv:1904.05811.

Han, Z., Chen, P., Ma, Y., Tresp, V., 2021. Explainable subgraph reasoning for forecasting on temporal knowledge graphs. In: International Conference on Learning Representations.

Karami, H., Darvish, K., Mastrogiovanni, F., 2020. A task allocation approach for human–robot collaboration in product defects inspection scenarios. In: 2020 29th IEEE International Conference on Robot and Human Interactive Communication (RO-MAN). IEEE, pp. 1127–1134.

Kovalenko, I., Balta, E.C., Qamsane, Y., Koman, P.D., Zhug, X., Lin, Y., Tilbury, D.M., Mao, Z.M., Barton, K., 2021. Developing the workforce for next-generation smart manufacturing systems: a multidisciplinary research team approach. Smart and Sustainable Manufacturing Systems 5, 4–24.

Li, S., Zheng, P., Pang, S., Wang, X.V., Wang, L., 2023. Self-organising multiple human–robot collaboration: A temporal subgraph reasoning-based method. Journal of Manufacturing Systems 68, 304–312.

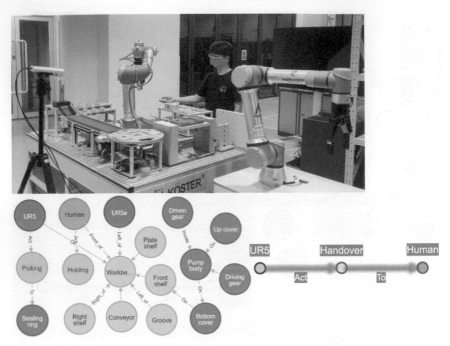

Figure 6.18 The gearbox assembly task with multiple HRC operation sequences – step 8.

Liau, Y.Y., Ryu, K., 2022. Genetic algorithm-based task allocation in multiple modes of human–robot collaboration systems with two cobots. The International Journal of Advanced Manufacturing Technology 119, 7291–7309.

Liu, X., Zheng, L., Shuai, J., Zhang, R., Li, Y., 2020. Data-driven and AR assisted intelligent collaborative assembly system for large-scale complex products. Procedia CIRP 93, 1049–1054.

Lu, Y., Zheng, H., Chand, S., Xia, W., Liu, Z., Xu, X., Wang, L., Qin, Z., Bao, J., 2022. Outlook on human-centric manufacturing towards Industry 5.0. Journal of Manufacturing Systems 62, 612–627.

Veličković, P., Cucurull, G., Casanova, A., Romero, A., Lio, P., Bengio, Y., 2017. Graph attention networks. arXiv:1710.10903.

Yang, J., Lu, J., Lee, S., Batra, D., Parikh, D., 2018. Graph R-CNN for scene graph generation. In: Proceedings of the European Conference on Computer Vision (ECCV), pp. 670–685.

CHAPTER 7

Deployment roadmap of proactive human–robot collaboration

This chapter presents a stepwise procedure for the development of Proactive HRC systems comprising four key modules: scene perception, knowledge representation, decision making, and collaborative control. For each module, we provide a comprehensive research roadmap of related technologies and offer an advanced algorithm as a feasible solution. The perception module is dedicated to perceiving the human–robot–workspace environment, as detailed in Section 7.1. Meanwhile, knowledge representation focuses on acquiring semantic knowledge of manufacturing tasks and transferring human expertise to robots for cognitive inference, as illustrated in Section 7.2. In Section 7.3, we delve into the decision-making module, which empowers the HRC system to make intelligent decisions for optimized trajectory planning and human information support, adapting to changing environmental conditions. Additionally, Section 7.4 provides an overview of various algorithms for robot collaborative control at the operational level. These four aspects have witnessed the widespread adoption of cutting-edge cognitive computing techniques such as DL, RL, TL, Large Language Model (LLM), etc., resulting in significant enhancements to Proactive HRC system performance.

7.1 Scene perception

Scene perception is the first step to developing a Proactive HRC system, which perceives the states of humans, robots and surrounding environments. Through the content awareness characteristic of Proactive HRC, both robots and humans gain the ability to intuitively understand the teammate's current status, locate the ongoing work area, identify missing components, and determine the appropriate tools for the task. This allows them to proactively decide on next collaborative actions.

Proactive Human–Robot Collaboration Toward Human-Centric Smart Manufacturing
https://doi.org/10.1016/B978-0-44-313943-7.00014-4

Table 7.1 Typical research efforts on the perception of human position, pose, motion, and activity.

Objective	Method	Reference
Hand gesture recognition	LSTM and CNN	Simao et al. (2019)
Limb pose recognition	LDA-based classifier	Lanini et al. (2018)
Voice recognition	DTW and user-dependent dictionary	Birch et al. (2021)
EEG-based arm movement recognition	LSTM and CNN	Buerkle et al. (2021b)
EMG-based human lower limb motion estimation	LSTM and adaptive robust iterative learning control	He et al. (2022)
Human activity recognition	Relation history image extraction	Gori et al. (2016)
Human activity recognition	Spatio-temporal joint based CNN	Abdelkawy et al. (2020)
Human group activity recognition	Multisensory data and Laplacian embedding	Lu et al. (2021)

7.1.1 Human operator perception

In Proactive HRC systems, human operator perception involves capturing various aspects of a worker's status, including their position, posture, movement, and activities. These can be observed through diverse means such as gestures, voice commands, biological signals, and motion sequences, as detailed in Table 7.1. Simao et al. (2019) utilized wearable sensors to recognize both static and dynamic human gestures. These gestures were then translated into commands for the robot, enabling reactions such as movement cessation, end-effector rotation, and gripper manipulation. Similarly, Lanini et al. (2018) monitored changes in the human upper body's actions through force sensor signals, using them as input commands for the robot. Voice control was also explored by Birch et al. (2021) in machine hole drilling, with a system that translated voice inputs into robot movements. To enhance human safety, Buerkle et al. (2021b) employed mobile electroencephalogram (EEG) signals to analyze an operator's motion intentions before execution. The analysis results provided early warnings to the robot, allowing it to proactively adjust its motion paths to prevent collisions. Additionally, He et al. (2022) estimated human lower limb motion intentions using surface electromyography (EMG) signals and developed a dynamic model for exoskeleton robot control. Recognizing that human motion sequences contain valuable operation intentions, Gori et al. (2016) explored an activity recognition approach, wherein humans simultaneously and sequentially performed different types of actions, going beyond short-term operator motions. Building on this foundation, Abdelkawy et al. (2020) investigated the semantic context between human activities and environmental events. Lu et al. (2021) subsequently expanded this exploration to include group activity recognition based on multisensory input data. Leveraging knowledge from these detected human activities, robots can make suitable decisions to facilitate and assist human operators in their tasks.

On the other hand, monitoring the physical and mental status of human operators, including fatigue and stress, is essential during Proactive HRC task execution. This aspect of human well-being has been extensively investigated, as summarized in Table 7.2, with a focus on body poses, EMG signals, and EEG states. For example, Kim et al. (2017) introduced a statically equivalent serial chain model to assess the load on human joints and adjust robot trajectories during collaboration tasks. Peternel et al. (2018) employed EMG sensors to measure muscle activity and fatigue, adapting robot execution speed and frequency to allow human recovery during different phases of wood sawing tasks. Peternel et al. (2019) addressed muscle fatigue management by redistributing external forces among specific muscle groups, enabling the robot to selectively assist tired muscles. This approach maximized human fatigue-related endurance and improved ergonomics for co-manipulation tasks. In an HRC polishing scenario, El Makrini et al. (2022) employed a postural optimization technique to reduce human workload. They used a graphical interface to identify nonergonomic postures and instructed the robot to adjust the pose of a comanipulated component accordingly. Lorenzini et al. (2019) measured whole-body overloading torques via EMG signals to evaluate cumulative fatigue induced by light payloads. The robot continuously adjusted co-manipulated postures to enable ergonomic repetitive tasks. To alleviate mental pressure during unexpected situations, Buerkle et al. (2021a) employed a mobile EEG sensor to detect potential emergencies in HRC tasks. This allowed the robot to react swiftly to prevent harm when incidents like workpiece drops or malfunctions occurred. Aldini et al. (2019) assessed cognitive conflict experienced by users when mechanical resistance opposed their motions during physical Human–Robot Interaction (HRI). The results informed impedance control adjustments to enhance operator comfort and safety. In order to enable HRC systems to adapt to the unique and dynamic nature of human behaviors, Buerkle et al. (2022) evaluated workload using objective, subjective, and physiological metrics in

Table 7.2 Typical research efforts on the analysis of human fatigue and pressure status.

Objective	Method	Reference
Body joint overload estimation	Statically equivalent serial chain	Kim et al. (2017)
Human muscle fatigue estimation with EMG	Two-order system for fatigue estimation and robot control	Peternel et al. (2018)
Selective muscle fatigue management	Force prediction, muscle fatigue model and impedance controller	Peternel et al. (2019)
Cumulative effect of the overloading fatigue	Whole-body fatigue estimation used RC circuit model	Lorenzini et al. (2019)
Postural optimization of neck, trunk and leg	Human body model using virtual kinematics chains (springs and dampers)	El Makrini et al. (2022)
Cognitive load between human–robot exchanged forces	Independent component analysis and admittance control	Aldini et al. (2019)
Human mental pressure detection	Decision tree model and continuous wavelet transform peak counting	Buerkle et al. (2021a)
Objective, subjective, and physiological assessment	ML and RL	Buerkle et al. (2022)

both manual and collaborative assembly tasks. These metrics include EEG signals, the NASA task load index, task completion time, and the number of errors or assistance requests.

In this way, human operator perception can enable robots to detect movements, understand operation intentions, and analyze fatigue and stress levels, for intelligent robot decision-making.

7.1.2 Robot status perception

In Proactive HRC systems, robots serve two primary functions: supporting humans or actively executing tasks. These motions are controlled through various parameters, such as safety measures, intuitive programming, and adaptive path planning,. In this context, the research endeavors aimed at improving the readability of robot operations are summarized in Table 7.3.

Table 7.3 Typical research efforts on readable robot status.

Objective	Method	Reference
Robot position	Robot light skin	Tang et al. (2019)
Robot legibility cue	Projected arrows and flashing lights	Hetherington et al. (2021)
Robot anticipatory motion	Facial gestures and motions	Gielniak and Thomaz (2011)
Robot trajectory	VR & DT	Oyekan et al. (2019)
Robot state	AR system	Wang et al. (2020)
Robot state preview	Zoomorphic gestures	Sauer et al. (2021)

For instance, in Tang et al. (2019), a robot's current position was indicated using illuminated skins, attracting human attention to ensure safety during manufacturing operations. Hetherington et al. (2021) designed mobile robots equipped with projected arrows and flashing lights to convey information about their paths and goals. To expedite human understanding of robot intent, Gielniak and Thomaz (2011) enhanced robots with facial gestures and human-like motions. Leveraging advancements in VR and DT technologies, Oyekan et al. (2019) employed digital spaces to represent trajectory information of physical robots. These visual cues offered an intuitive understanding of robot status within noisy factory environments. Furthermore, Wang et al. (2020) developed an AR-based HRC system to present robot states in a virtual-reality fusion environment, which combined precise robot control with intuitive human information support. To address more complex robot states, such as task progress, Sauer et al. (2021) introduced gesture-based zoomorphic methods, and assessed user preferences regarding communication styles (e.g., attractiveness, joy, intuitiveness). Collectively, these research efforts aimed to provide human-readable indicators of robot operational goals, thereby enhancing human adaptability to robots in Proactive HRC scenarios.

7.1.3 Workspace perception

To facilitate smooth and efficient Proactive HRC operations, it is crucial to accurately determine the positions and orientations of mechanical components within the workspace. As illustrated in Table 7.4, various approaches have been explored to achieve this goal. For localization precision, Lee et al. (2018) introduced object segmentation techniques, specifically the Canny edge detector, to extract detailed shape information from work-in-progress parts during electric motor assembly tasks. Rosenberger et al.

Table 7.4 Typical research efforts on surrounding workspace perception.

Objective	Method	Key Elements	Reference
Object segmentation	Canny edge detector	Shape information of part	Lee et al. (2018)
Object detection	YOLOv3	Industrial components in a cluttered environment	Rosenberger et al. (2020)
2D pose estimation	Shaver handles	Mapping 3D CAD model to images	Tsarouchi et al. (2016)
6D pose estimation	OBB & ICP	Point cloud construction and post refinement	Franceschi et al. (2020)
SG	GCN & RNN	Graph map of preception results	Moon and Lee (2018)
2D map	3D CNN	Occupancy grids-based robot team position	Dias et al. (2020)
3D representation	OctMap & PoseNet	3D occupancy status of workspaces	Liu and Wang (2021)

(2020) employed a deep CNN series model (YOLOv3) to detect industrial components amidst cluttered backgrounds. Tsarouchi et al. (2016) estimated the pose of shaver handles based on predefined CAD models, enabling robotic pick-and-place operations in production lines. Furthermore, for precise 6D pose estimation of mechanical components in manufacturing tasks, Franceschi et al. (2020) employed Oriented Bounding Box (OBB) for rough results and subsequently utilized the Iterative Closest Point (ICP) algorithm to refine them. Moon and Lee (2018) employed a GCN-based 3D semantic graph map to construct a detailed scene description of the surrounding environment. Dias et al. (2020) explored the use of an occupancy grid to represent the positions of a team of robots, which also served as an interactive interface for controlling robot sequences. Additionally, Liu and Wang (2021) created a 3D occupancy status map of the HRC workspace using OctMap to enable active collision avoidance by the robot. Perceiving the shared workspace empowers the robot to gather information about the positions and shapes of various objects, enabling it to adapt to changes in the environment, including both dynamic and static obstacles.

7.1.4 A human digital twin example

In accordance with the definition in Fan et al. (2023), the Human Digital Twin (HDT) emerges as a pivotal concept, representing a virtual counterpart of a human worker designed to enhance collaboration experiences. Within this framework, the modeling of the HDT stands as an effective strategy to interpret vital attributes pertaining to human involvement in Proactive HRC. This comprehensive HDT model seamlessly fuses diverse sources of data, creating an integrated representation for the description, prediction, and visualization of an individual or a specific human profile within human–robot teamwork. The primary objective aims to optimize worker well-being while concurrently ensuring that the collaboration between humans and robots is both efficient and secure.

Within the realm of human perception in Proactive HRC, prior investigations have focused on two essential aspects: 1) the recognition of human body skeletal structures to facilitate active collision avoidance, and 2) the prediction of human action intentions to inform robotic decision-making processes. A notable example in this regard is the work of Wang et al. (2018a), who introduced a deep learning-driven model adept at identifying human assembly actions, particularly during a collaborative engine disassembly task involving human–robot cooperation. Similarly, the study conducted by Liu and Wang (2021), featured the application of PoseNet for the recognition of human body skeletons and employed OctoMap as the vehicle for environment representation within the human–robot interaction context. This combination of technologies enabled enhanced context awareness and the avoidance of collisions. Furthermore, the research carried out by El Makrini et al. (2022) delved into the application of the skeleton tracking algorithm with the Kinect camera. This algorithm was coupled with the Rapid Entire Body Assessment (REBA) ergonomic score proposed by Hignett and McAtamney (2000) to optimize the position of the end-effector in HRC tasks. Nonetheless, it is noted that these approaches have displayed certain limitations, often resulting in relatively coarse representations of the human operator and falling short in achieving fidelity and recognition precision of human modeling. To address these shortcomings, a novel and comprehensive HDT modeling scheme has been proposed and is depicted in Fig. 7.1. This model is structured around three core components: 1) fine-grained human pose reconstruction, 2) spatio-temporal human action recognition and ergonomic evaluation, and 3) multimodal intelligence-based human fatigue and pressure estimation.

Fine-grained human posture reconstruction is the first part of the HDT. A deep learning model that can simultaneously reconstruct the fine-grained 3D dense mesh and skeleton joints of the human body is proposed. Concretely, the RGB-D images of the human operator will be processed by

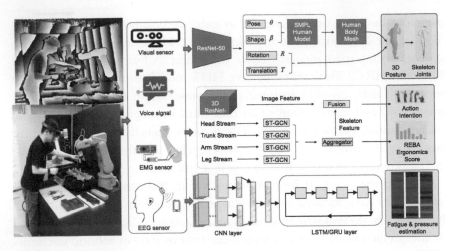

Figure 7.1 Human digital twin for human perception in Proactive HRC. (Adapted from (Zheng et al., 2023).)

an ResNet-50 backbone network to extract the geometric features, which is then utilized to regress the pose parameters $\theta \in \mathbb{R}^{3 \times K}$, shape parameters $\beta \in \mathbb{R}^{M}$, 3D rotation $R \in \mathbb{R}^{3 \times 3}$, and 3D translation $T \in \mathbb{R}^{3}$. The pose and shape parameters are subsequently sent to the SMPL (Skinned MultiPerson Linear model) human body model (Loper et al., 2023) – a differentiable function that outputs a triangulated mesh $M(\theta, \beta) \in \mathbb{R}^{3 \times N}$. The adoption of SMPL model can simplify the reconstruction process to achieve real time performance by relying on a template human body mesh as a prior which will be bended and stretched to the target human pose according to the estimated pose and shape parameters. The predicted global 3D rotation and translation will be applied to obtain the correctly posed human body mesh. The 3D skeleton points $X(\theta, \beta) \in \mathbb{R}^{3 \times K}$ are further obtained by linear regression from the final mesh vertices. Then, to refine hand posture reconstruction of human operators, we additionally adopt the MANO (hand Model with Articulated and Nonrigid deformations) to enhance the SMPL model following the practice in Romero et al. (2022). Once the fine-grained human mesh is reconstructed, it can represent the precise geometric occupancy of the human body, which can substantially reduce the perception uncertainty of the human body during HRC.

The spatio-temporal recognition of human actions and the evaluation of ergonomic factors represent another components of HDT development, serving as semantic knowledge. This phase is vital for the comprehensive construction of HDT, particularly when dealing with high-level reasoning and optimization of robotic movements for human-centric concerns. Within this module, two primary data sources are considered: the RGB video stream and the associated skeleton stream. The RGB stream undergoes processing via a 3D ResNet-50, an approach that effectively handles spatio-temporal features in a unified convolution structure. This process extracts temporal image features, which contribute to the subsequent stages. Simultaneously, the skeleton stream is divided into four distinct branches: the head, trunk, arm, and leg branches. Each of these branches is processed through a specialized ST-GCN network, extracting the spatio-temporal features of local body parts. To create a global skeleton feature based on these branches, an aggregator network is used to fuse the features derived from the individual body segments. This fused skeleton feature is then integrated with the extracted image feature, to discern the specific type of human behavior currently in progress. Furthermore, automatic evaluation of ergonomic risks pertaining to the human body is importance within the context of robotic actions, particularly when aimed at reducing the occurrence of occupational diseases in an environment that considers human-centric principles. In this section, the REBA ergonomic assessment tool is introduced, which effectively rates the risk of musculoskeletal disorders associated with a particular body posture. These ratings are assigned on a scale ranging from 1 to 15. While it is conceivable to compute the REBA scores directly from the body joint data, the nondifferentiable nature of this process renders it impractical to be directly integrated into a gradient-based optimization model. Consequently, an alternative approach has been adopted, wherein the REBA score is regressed through the involvement of an additional neural network branch. This branch mirrors the design used in the action branch of the final fully connected layer. To supervise the training process effectively, a Smooth L1 loss function (Ren et al., 2015) is deployed, due to its smoother gradient transition at 0. This feature can improve regression performance. The outcome of this approach notably improve accuracy performance in ergonomic score predictions.

To enhance the overall safety and well-being of individuals engaged in close and repetitive work, it is imperative to assess and address human fatigue and pressure levels effectively. This multimodal approach incorporates the measurement and analysis of both EMG and EEG signals. To ensure the integrity and accuracy of these measurements, filtering processes are employed to eliminate undesired deviation and noise of the EMG and EEG signals. Subsequently, a 1D CNN model is utilized. It can extract features from the signals for in-depth fatigue analysis, while estimating pressure levels. Incorporating LSTM, the model can refine the extracted information obtained from the sequential data. In this way, this model can continuously update the cumulative measures of human fatigue and pressure. Then, the integrated features are fused into the HDT model, which can monitor and optimize human-centric concerns in real-time.

7.1.5 Multi-granularity scene segmentation and 3D dense hand–object pose estimation

In close collaboration between humans and robots, the safety during interactive operation relies on scene segmentation of the shared workspace. Besides, precise 3D pose estimation of industrial components handover in the human hand is a precondition for collaborative execution. This scene perception approach goes beyond the mere monitoring of human states; it dynamically separates overlapping and occluded workpieces within the complex environment, facilitating fluent and proactive collaboration.

Past explorations focused on the task of semantic segmentation utilizing RGB-D information. For instance, in Hu et al. (2019), a 2D map generated through SLAM was employed to represent the global environment. Additionally, the Mask R-CNN model was adopted to achieve instance-level scene segmentation, for robotic observations of local environments. Similarly, Rozenberszki et al. (2021) introduced a pipeline for constructing 3D semantic maps using human-generated data and subsequently transferring the semantic labels to robotic agents. Although these works tackled various pixel-level prediction tasks, such as edge mapping, surface normal, and object part segmentation, they cannot achieve the implementation of scene segmentation across multiple granularities. The reliance on single-level semantics impedes scene construction across diverse activities within HRC.

At the operation level, it is necessary for the robot to attain precise posture information of the target objects overlapping with human hands, especially in real-time human–object interactions. To address this need, Hoang and Jo (2018) introduced the CMNN (Cascade of Multiple Neural Network) for the estimation of 3D human body poses from a single RGB image, expressed as 3D body joints. Similarly, Lin et al. (2018) concentrated on pose estimation for vehicle parts in automated spray painting. In this scenario, 3D point cloud data was processed through the Iterative Closest Point (ICP) and Genetic Algorithm (GA) to attain the final pose estimation results. However, these works only pay attention to ideal scenarios, where objects were presented in isolation without interference from or interactions with humans.

For Proactive HRC targets at realistic and complex scenarios, a combined approach for scene perception proves to be applicable. This approach contains two components: multi-granularity scene segmentation and integrated 3D hand–object dense pose estimation, as shown in Fig. 7.2.

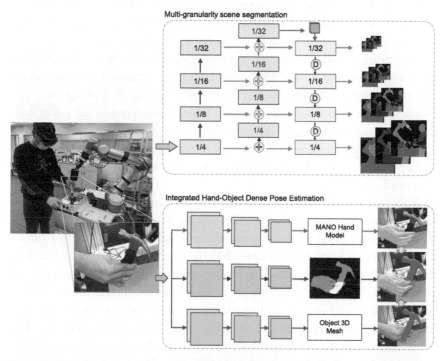

Figure 7.2 Multi-granularity scene segmentation and 3D dense hand–object pose estimation in Proactive HRC. (Adapted from Fan et al. (2022a) and Fan et al. (2022b).)

To enhance the adaptability and perception capabilities of collaborative robots, the scene segmentation method for Proactive HRC is structured into three levels: the area level, entity level, and part level. Each of these levels focuses on a specific aspect during the collaboration. The area level is primarily employed to address tasks of coarse-grained environment construction, particularly for robotic navigation within a range of work areas. It allows understanding of the overall workspace, thus robots can navigate and position themselves optimally. The entity level aligns with the conventional semantic segmentation criteria, focusing on the segmentation of different entities to diverse semantic categories. This level is crucial for identifying and distinguishing between objects and entities in the robot's vicinity. The part level takes a more granular approach, breaking down these entities into their constituent parts, for potential interactions with robotic end effectors. It enables a detailed understanding of how different components of an entity can be manipulated with during the collaborative process. The multi-granularity scene segmentation model is in the encoder–decoder structure. The encoder is equipped with two branches of networks, derived from the ConvNext model (Liu et al., 2022c) for processing RGB and depth information. The decoder segment derives from the ESANet framework (Seichter et al., 2021) and is further extended to incorporate features of multi-scale refinement and segmentation. Following two upsampling stages in the decoder, the final segmentation results are generated. The three branches of the multi-granularity segmentation approach present similar network structures. The primary distinction among them is the final output channels, which correspond to the different granularity levels and associated categories.

Followed by 3D dense reconstruction of hand–object poses, the approach even can deal with partially obscured observations in Proactive HRC activities. The model comprises three key components: (a) mask-guided attentive feature extraction, (b) hand–object dense pose estimation, and (c) occlusion awareness. The process commences with the capture of a monocular RGB image, denoted as I_o, within the HRC scene. Subsequently, the model detects and identifies the hand–object interaction area and recognizes the object type, leveraging an efficient detection model. For this purpose, a YOLOv3-tiny object detector (Redmon and Farhadi, 2018) is employed due to its computational efficiency. The hand–object interaction area, designated as I_{ho} and residing in $I_{ho} \in \mathbb{R}^{3 \times H \times W}$, is then extracted from the original image I_o and serves as the input to our model. The feature extraction component utilizes a backbone network, structured based on ResNet50 (He et al., 2016), to extract hand and object features, guided by binary masks associated with the hand and object. Subsequently, in the hand–object pose estimation section, the extracted feature vectors undergo processing through multiple FC layers. This processing results in the prediction of pose parameters, including hand translation, pose, shape, object rotation, and object translation. These parameters are then applied to both the 3D object model and the MANO hand model (Romero et al., 2022) to generate 3D geometric reconstructions. The final component, occlusion awareness, can handle situations where parts of the hand or object are obscured. To achieve this, a ternary mask is generated through the projection of 3D reconstructions onto the 2D image plane via differentiable rendering (Kato et al., 2018). Simultaneously, intermediate feature maps from the feature extraction phase are employed to establish a subnetwork following the principles of an Feature Pyramid Network (FPN)-like architecture (Lin et al., 2017). This subnetwork predicts the ternary mask in a segmentation manner. To ensure the accuracy and consistency of the reconstructions despite potential occlusion, the model calculates the discrepancy between the predicted ternary mask and the mask generated through rendering. This step minimizes reconstruction errors arising from occluded regions in the hand–object interaction.

This methodology provides the Proactive HRC system with the capability to perform two essential tasks: Firstly, it can adeptly segment an array of industrial components within the shared workspace, facilitating the precise identification of these components in an environment where they may overlap. Secondly, it enables the system to discern intricate details of the interactions between humans and these objects, providing knowledge of the poses and postures assumed by humans in relation to the objects they are handling during the execution of tasks.

7.2 Knowledge representation

Semantic knowledge extraction from HRC scene perception results enhances task comprehension, facilitates the learning of human behaviors, and enables the reception of commands through human brainwave analysis. This rich knowledge representation is vital for in task parsing, imitation learning, and brain-controlled robots.

7.2.1 Task parsing and understanding

In Proactive HRC, the objective of task parsing and understanding is to infer operational requirements and task objectives. As demonstrated in Table 7.5, several approaches have been developed for this purpose. Ahn et al. (2018) introduced Text2Pickup networks, enabling the HRC system to generate inquiries when confronted with ambiguous task instructions. This bi-directional communication allows robots to grasp the task's intent and select the necessary objects from the workspace. Subsequently, Tan et al. (2020) employed Visual Question Answering (VQA)-based methodologies to achieve task comprehension in HRC by encoding information

Table 7.5 Typical research efforts on task parsing and understanding.

Objective	Method	Reference
Understanding of ambiguous task instructions	Text2Pickup Network for object detection and language command understanding, question generation network for human commands feedback	Ahn et al. (2018)
VQA for task instructions	Object, detection, and gesture recognition, symbolic reasoning for answer and instruction	Tan et al. (2020)
Scene understanding for safety analysis	Mask R-CNN object detection, multilevel SG	Riaz et al. (2020)
Inferring semantic properties of the world	Probabilistic representation for semantic knowledge, Bayesian formulation for incremental estimation	Arkin et al. (2020)
Hierarchical task analysis	Part dependent task complexity and task functional structure	Mateus et al. (2019)
Hierarchical subtask plan	Bayesian inference for trajectory prediction, DTW distances based plan recognition, a planner for robot motions	Cheng et al. (2020)
Explicit temporal constraints of task sequences	Maximum entropy inverse optimal control for human goals, partially observable Markov decision process for robot goals	Pulikottil et al. (2021)

from speech, gestures, and visual cues. To capture contextual information within the current environment, Riaz et al. (2020) employed a multi-level scene description neural network to predict the SG in a warehouse setting. The SG proves can forecast risk of task operations and avoiding unsafe situations. Leveraging prior factual knowledge, Arkin et al. (2020) learned task-relevant instructions from both human linguistic descriptions and the robot's physical interactions with the surroundings. These inferred semantic knowledge of the environment enabled the robot to rectify errors in task instructions. Mateus et al. (2019) introduced a task decomposition structure that generated multiple task options while considering collaborative modes and spatial allocation of resources and components. Cheng et al. (2020) decomposed tasks into hierarchical and temporal subtasks and plans. They explored the hierarchical relationships among action sequences required to accomplish the task. Furthermore, Pulikottil et al. (2021) inferred knowledge regarding temporal constraints based on human and robot operational goals. With knowledge of task constraints, the Proactive HRC system could plan robot motions for fluent collaboration. For instance, it ensured that humans did not manipulate objects handled by the robot until the robot had completed its task.

7.2.2 Imitation learning and human skill transfer

In Proactive HRC, the robot learning processes involving the imitation of human behaviors and the transfer of human skills, are illustrated in Table 7.6. Various approaches have been developed to achieve these objectives. Wang et al. (2018b) introduced a teaching–learning–collaboration model enabling robots to learn from human demonstrations. The robot could acquire suitable motion trajectories from human instructions and operation sequences to perform new tasks. Sun et al. (2020) employed a learning from demonstrations framework to enable robots to assist humans effectively. Meanwhile, Sasagawa et al. (2020) proposed bilateral control-based imitation learning, allowing robots to adjust force and motion speed when mimicking complex human motions. For the transfer of human welding skills, Wang et al. (2019) utilized VR hardware for teleoperating a 6-DoF robot. The VR platform facilitated the intuitive transfer of new operation trajectories from human welders to robot manipulations. To enhance user-friendly robot programming, Macchini et al. (2019) presented a user-adapted interface through personalized body–machine mapping, enabling the natural teleoperation of mobile robots with high accuracy. Losey and O'Malley (2019) employed Bayesian inference to learn interaction strategies between robots and end-users, enabling personalized robot teaching. This approach could create personalized HRI that accommodated human uncertainty during interactions and reduced the ambiguity in fixed, predefined interactive strategies. For precise robot motion during kinesthetic teaching, Chen et al. (2020) recognized human actions and extracted action features sequentially from velocity, force torque data, and gripper information. These parameters were then reconstructed into robot trajectories, ensuring a high level of motion accuracy (2.37 mm). This personalized configuration allowed for intuitive control by humans and facilitated the transfer of their skills to complex robotic systems. Furthermore, to transfer human expertise to robotic systems for grinding of complex surfaces, Marullo et al. (2020) proposed a decentralization control strategy based on contact and impedance forces. Humans utilized force feedback to adapt the robot's path trajectory to slightly deformed surfaces.

Table 7.6 Typical research efforts on imitation learning and human skill transfer.

Objective	Method	Reference
Learning from demonstration	Speech recognition and learning of task-based knowledge	Wang et al. (2018b)
Learning from demonstration	A dual-input DL algorithm, online automated data labeling	Sun et al. (2020)
Human skills imitation	4ch bilateral control for human–robot cooperative execution in terms of force information and fast motion	Sasagawa et al. (2020)
Welders' operation skill transfer	VR head-mounted display, motion-tracked handle controller, welder operation prediction	Wang et al. (2019)
Personalized body-machine interface	CCA and SVR for personalized motion synergy identification	Macchini et al. (2019)
Personalized HRI	Bayesian inference of human interaction strategy, inverse RL simulation for robot motion	Losey and O'Malley (2019)
Kinesthetic teaching of robot arm tasks	Human motion recognition and segmentation, a hybrid sensing interface for motion feature recording	Chen et al. (2020)
Path trajectory for surface sanding	A decentralized control strategy based on impedance force	Marullo et al. (2020)

7.2.3 Brain-controlled robot

The process of translating human brainwave patterns into robot control commands relies on knowledge representation learning algorithms, as detailed in Table 7.7. For robust robot control in noisy environments, Wang et al. (2021) integrated brainwave signals with function block commands, enabling adaptive robot actions in engine assembly. This approach proved particularly useful when workers were occupied with other concurrently handling tasks. To establish natural communication between human physiological responses and optimal robot planning, Liu et al. (2022b) employed wearable EEG and EMG bio-sensors to create a physiological communication interface. This interface provided essential human physiological awareness as a prerequisite for making logical decisions regarding robotic control and manipulations. Furthermore, Liu et al. (2022a) combined multi-modal commands derived from brainwaves, gestures, and voice to accurately trans-

Table 7.7 Typical research efforts on brain-controlled robots.

Objective	Method	Reference
Brainwave control	Wavelet transform and TL algorithms	Wang et al. (2021)
Human physiological response and robot optimal planning	Wearable EEG and EMG biosensors, ANN-based classifier, robotic control system	Liu et al. (2022b)
Translation of brainwave command phrases into robot commands	Brainwaves, gestures and voice commands, a DL algorithm for command classification, function block for robot control	Liu et al. (2022a)

late instructions for the robot. These brain-controlled methods resulted in clear operation schedules and minimized confusion during collaboration tasks involving both humans and robots.

7.2.4 An example of knowledge graph representation for HRC settings

The KG method is an effective way to encapsulate knowledge representation within the dynamic settings of HRC. By employing KG, it enhances the cognitive intelligence in Proactive HRC systems. Within the KG structure, nodes can be utilized to convey a diverse array of components such as humans, robots, workpieces, tasks, and environmental factors, all of which play pivotal roles in Proactive HRC scenarios. Simultaneously, the edges interlinking these nodes within the KG serve to encode the relationships and connections that exist among these constituent elements.

In previous studies, efforts were made to extract semantic knowledge within the context of HRC. For example, Rahman et al. (2016) introduced an HRC framework to determine the appropriate sensing mode (human or robot) for assembly parts detection. This decision was made based on evaluation of the confidence levels and associated costs of observations, employing a regret-based Bayesian decision-making approach. Additionally, Riley and Sridharan (2019) explored explanatory VQA methods within HRC. This approach involved the integration of CNN, Recurrent Neural Network (RNN), logical reasoning, and inductive learning. Nevertheless, these approaches were confined to reasoning within specific knowledge domains within the execution loop, failing to establish a comprehensive knowledge framework that spanned the entirety of the HRC process.

To learn a comprehensive representation of the Proactive HRC setting, a domain-specific KG can be created in a hierarchical and systematic manner, which encompasses three distinct layers, as depicted in Fig. 7.3. These layers are designed to capture nodes and features related to HRC scenarios. The HRC Task Layer, at the highest level of the hierarchy, is used for the identification and sequencing of incoming tasks. It outlines the sequential order of tasks, describes their interdependencies, and defines the stages of execution. This layer is pivotal for task-oriented configuration searches, ensuring that tasks are carried out in a well-defined sequence for safety concerns. The Allocation Layer plays a crucial role in associating humans and robots with various HRC subtasks, establishing intra-layer relations to indicate which individuals are involved in specific subtasks. Moreover, it defines the relationships between different participants, whether they are human–human, robot–robot, or human–robot interactions. These relationships help delineate the roles of participants within each subtask and the role of their collaboration. The Configuration Layer, at the lowest level, delves into the specifics of actions and movements, describing their relationships using terms like "part of", "has", "to do", and so on. This layer provides detailed configurations for both humans and robots, enabling them to fulfill tasks efficiently. For humans, it includes aspects such as workpiece selection, positioning, and instructions. The schema also encompasses environment and workpiece nodes, which provide relevant configuration details. Additionally, the robot's missions, including safety and path planning, are associated with their respective objects. By employing this HRC-based KG, HRC subtasks can be dynamically allocated to humans, robots, or a combination of both, based on the selected collaboration modes, fostering effective and safe task execution.

7.3 Decision making

In Proactive HRC, dynamic decision-making is pivotal. It involves making appropriate task plans, offering interactive support to humans, and optimizing robot operations.

7.3.1 Task planning and role allocation

In Proactive HRC, task planning involves allocating various operational responsibilities to human and robotic agents based on their respective capabilities. As detailed in Table 7.8, multiple strategies have been devised to

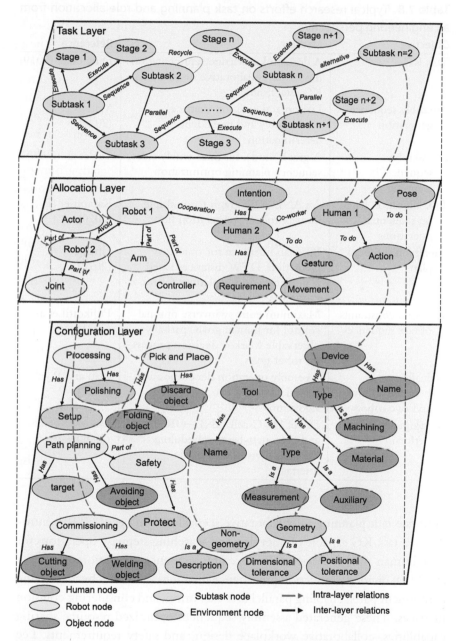

Figure 7.3 Knowledge graph for knowledge representation in Proactive HRC. (Adapted from Zheng et al. (2022).)

Table 7.8 Typical research efforts on task planning and role allocation from an engineering perspective.

Objective	Method	Reference
Task KG	A domain knowledge representation model for collaborative disassembly, KG establishment	Ding et al. (2019)
Assembly sequence generation	Liaison and collision matrices, subassembly and precedence determination	Mateus et al. (2020)
Disassembly sequence planning	A graph model for disassembly rules, sequence planning optimization	Lee et al. (2020)
Task allocation based on sequential operations	An AND–OR graph for multiple human–robots cooperation flow	Karami et al. (2020)
Hierarchical subtask plan	Bayesian inference for trajectory prediction, DTW distances based plan recognition, a planner for robot motions	Cheng et al. (2020)
Temporal constraints of task sequences	Maximum entropy inverse optimal control for human goals, partially observable Markov decision process for robot goals	Pulikottil et al. (2021)
Task allocation between a human and two robots	Assembly operation decomposition, analytic network process, genetic algorithm	Liau and Ryu (2022)
Next operations to perform	DT of HRC cell, AND–OR tree and Petri nets-based scheduling algorithm, visual and tactile interface	Maderna et al. (2022)

optimize task planning in collaborative scenarios. Ding et al. (2019) introduced a task KG capable of querying and searching stepwise operations for both human and robot agents. For assembly sequence generation, Mateus et al. (2020) divided production precedence into sub-assemblies by identifying component parts, enabling parallel task execution, and considering collision matrices. These generated assembly sequences optimized resource sharing capabilities, collaborative workplace design, and safety requirements. Lee et al. (2020) introduced a graph model for disassembly sequence planning, adhering to disassembly rules, such as disassembly costs for robots and humans, various starting points, safety considerations for human operators, and feasible robot operations. For dynamic task allocation among a human operator, a mobile manipulator, and a dual-arm manipulator, Karami et al. (2020) utilized an AND–OR graph to represent concurrent and sequential operations within the team. Besides, Cheng et al. Cheng et al. (2020) decomposed tasks into hierarchical and temporal subtasks and plans, exploring the hierarchical relationships among action sequences to accomplish tasks efficiently. Their task planner leveraged task decomposition and human trajectory prediction to facilitate efficient subtask co-working in advance. Furthermore, Pulikottil et al. (2021) inferred task temporal constraints based on human and robot operational goals, allowing the Proactive HRC system to plan robot motions optimally. Lastly, Liau and Ryu (2022) developed a task allocation model for a single human and two robots, considering task characteristics and agent capabilities. This model decomposed the work process into functional actions based on component geometry, tolerance, and required force exertion. Maderna et al. (2022) developed an HRC system with flexible scheduling and tactile communication. Their scheduling algorithm considered variations in the duration of human tasks and potential robot faults to dynamically generate task allocation and sequences for human and robotic agents.

7.3.2 Dynamic interaction and support with human

The decision-making process in Proactive HRC facilitates dynamic interactions between humans and robots, coupled with domain knowledge support for humans, as depicted in Table 7.9. Over the years, various approaches have been developed to improve interaction and support in HRC. As early as 2005, Marín et al. (2005) pioneered a remote robot teleoperation system that made decisions based on human voice commands to control six-degree-of-freedom robot motions. These decisions ranged from navigating to a position with a permissible speed to grasping an object at a specific angle. Then, Wang et al. (2016) introduced an AR-based bare-hand interface to provide various types of guidance decisions to assembly operators at different phases of assembly tasks. This HRC system was found to be intuitive, user-friendly, and supportive for human with domain knowledge in assembly. For bi-directional decision exchange, Casalino et al. (2018) utilized wearable vibrotactile rings in HRC tasks. These rings enabled robots to understand the operator's upcoming actions, while the human operator received vibrotactile feedback. Likewise, Liu et al. (2020) employed AR devices to present intuitive assembly guidance instructions to humans during the assembly of large-scale complex products. Multiple operators in the

Table 7.9 Typical research efforts on dynamic interaction and knowledge support for human.

Objective	Method	Reference
Voice commands (grasp, position, rotation)	Web interface, remote programming based on networking protocols	Marín et al. (2005)
Multimodality assembly guidance	Bare-hand interface, AR guidance manager, user request interface	Wang et al. (2016)
Bi-directional information exchange	Human behavior estimation and wearable vibrotactile feedback	Casalino et al. (2018)
Assembly guidance of instructions	3D reconstruction of environments, multioperator interaction, human to machine interoperation	Liu et al. (2020)
Display of robot status, operator instructions, workspace changes	Projector and a wearable AR gear, depth-based workspace modeling, safety monitoring	Hietanen et al. (2020)
Decision and optimization from DT	AR and wearable devices, IIoT system, DT-based process design and optimization	Arkouli et al. (2021)
Procedural suggestion from DT	Knowledge-based DT, human motion modeling and simulation, action and attention recognition	Tuli et al. (2021)
Resource sharing along systems' life cycle	User interface, DT modeling and physical-digital space connection of HRC each phase	Malik and Brem (2021)
Manufacturing process management	Web interface, autonomous robot navigation, human position detection and tracking	Rey et al. (2021)

system could access and share assembly procedure information using AR helmets. Hietanen et al. (2020) utilized an interactive AR system to enable human operators to monitor real-time robot states and safety zone changes in the workspace, while allowing robots to receive on-the-fly instructions from users for task execution. In another approach, Arkouli et al. (2021) integrated decision-making and production scheduling optimization with real-time updates of the shop floor in DT environment. This information was delivered to human operators through AR tools, helping alleviate physical and cognitive stress. Tuli et al. (2021) developed a knowledge-based DT that categorized human activities into actions and predicted interaction re-

Table 7.10 Typical research efforts on robot trajectory re-planning considering task changes.

Objective	Method	Reference
Robot adaptation to a task goal	Bounded-memory adaptation model, mixed observability Markov decision process for robot actions	Nikolaidis et al. (2017)
Online adaptive path planning	A GMM-based algorithm, path re-planning closer to the task trajectory	Cheng et al. (2021)
Coordinated mobile robot movements	Connectivity-maintenance robot control, grid covering and path planning, wearable haptic feedback for humans	Aggravi et al. (2021)
Robot re-planning	Monte Carlo tree search, human semiautonomous teleoperation	Dalmasso et al. (2021)
Anytime robot path re-planning	PathSwitch algorithm, speed and separation monitoring, RRT	Tonola et al. (2021)

gions and objects. The DT-based HRC system could generate decisions regarding parts that might need assembly in the next step and provide suggestions to humans. Lastly, Malik and Brem (2021) utilized DT to construct collaborative production systems with high complexity. Their approach enabled decision-making and resource sharing for humans throughout the system's life cycle by modeling the physical–digital space connection for assembly tasks in each phase. Rey et al. (2021) introduced a web interface for managing decision information in manufacturing processes on the production line. Human operators created and updated orders through this interface, while AGV-based robots executed tasks like picking and transporting raw materials to mixing tanks.

7.3.3 Robot optimal trajectory planning

On one hand, robot task planning in Proactive HRC systems aims to allow robots to adapt and re-plan their trajectories to accommodate changing task requirements. Table 7.10 highlights various approaches to achieving effective robot re-planning. To facilitate appropriate robot action selection, Nikolaidis et al. (2017) introduced a probabilistic decision process that balanced the trade-off between gathering information on uncertain human behaviors and progressing toward task completion. The robot dynamically adjusted its motions for improved task execution. Cheng et al. (2021) proposed an online path planning algorithm for adaptive robot motion. This algorithm modified only the part of the path that encountered obstacles, while the remainder closely followed the original task trajectory. For adaptive robot navigation, Aggravi et al. (2021) presented a decentralized haptic-enabled connectivity-maintenance control framework for human–robot teams. Mobile rescue robots adapted their navigation movements based on the current connectivity level to facilitate easy human follow-up. Dalmasso et al. (2021) developed a multi-agent shared plan model, allowing robots to re-plan path motions according to human goals. Human operators had straightforward access to robot status information for further corrections and modifications. To achieve robot path re-planning at any time, Tonola et al. (2021) combined online path re-planning with speed optimization modulation. This re-planning algorithm enabled the robot to complete its task even in unexpected situations, such as the presence of humans or obstacles.

On the other hand, human-centric needs like human safety and ergonomic requirements require robots to adjust and optimize motions in Proactive HRC systems. This is exemplified in Table 7.11, where various strategies have been devised to address this consideration. For instance, to ensure robot manipulation aligned with human preferences, Granados et al. (2017) leveraged physical feedback from long-term interactions to enable the robot to comprehend human needs and adapt its behavior accordingly. Ergonomic risk prediction was applied to picking-and-placing tasks, enabling mobile robots to proactively assume high-risk actions or engage in repetitive medium-risk actions over extended durations. Ansari and Karayiannidis (2021) explored a task execution and control scheme that generated optimal robot forces to reduce human effort in cooperative object manipulation. The desired force was computed based on the human's desired trajectory for the identified task and then provided as input to the robot. The robot subsequently adjusted parameters to execute the task with the calculated force. Furthermore, Khatib et al. (2021) optimized robot kinematics to enable the robot to present a workpiece to humans with changing positions and orientations over time, ensuring ease of coordination for human workers.

7.3.4 An example of proactive HRC task planning based on graph embedding

Proactive task planning algorithms that utilize graph embedding offer the capability to dynamically learn both explicit and implicit knowledge re-

Table 7.11 Typical research efforts on robot optimal trajectory planning considering human safety and ergonomic requirements.

Objective	Method	Reference
Ergonomic robot control	Ergonomic risk prediction and pattern extraction of human actions	Parsa et al. (2019)
Robot actions with humans required force	Velocity identification of the human-performed task, robots' contribution to each degree of freedom of the task	Ansari and Karayiannidis (2021)
Time-varying poses desired by humans	Coordinated motion-based collision avoidance, saturation in the null space algorithm	Khatib et al. (2021)
Optimal robot sequence allocation	RL algorithms for task allocation, human fatigue model with adding noise	Zhang et al. (2022)

garding the progress of HRC tasks. This liberates existing HRC systems from the constraints of pre-defined instructions, resulting in improved adaptability in manufacturing tasks. By fusing historical collaborative knowledge and real-time on-site conditions, proactive task planning algorithms can anticipate task completion strategies, even when faced with unfamiliar or new tasks.

In previous research, various approaches have been explored for decision-making in HRC. Guo and Yang (2021) introduced a multi-variate Bayesian inference method to modele the dynamics of target specification in the HRC process. Yu et al. (2020) employed a chessboard-based simulation approach to mimic the assembly process, guiding decision-making in HRC. Zhang et al. (2022) used reinforcement learning methods to generate task operation sequences in HRC assembly procedures. Furthermore, Raatz et al. (2020) adopted a genetic algorithm to optimize task scheduling, factoring in capabilities and time constraints within HRC. However, these methods have limitations, as they focus primarily on explicit relationships in HRC, neglecting the implicit cues that specific human actions and robot states can have on task forecasting within HRC. This results in a failure to update and inform subsequent stages of task arrangement effectively.

With this context, we employ the EvolveGCN approach to facilitate proactive task planning for Proactive HRC, an illustration of which can be found in Fig. 7.4. To achieve this, the knowledge of the on-site scene perception should undergo aggregation via an Encoder model, expressed as $X_e = \sigma(W_h X_h + b_h)$. Here, $X_h = \{H, O, R, E\}$ signifies the consolidation of perceived scene results, with components such as human intention represented by H, object 6-DoF pose encapsulated in O, robot status denoted as R, and environmental parsing identified as E. On the other hand, X_e represents the Encoder's generated representation, while W_h and b_h correspond to the weight matrix and bias, respectively. The primary objective of the Encoder model is to translate the perception results into node-level embeddings. By accomplishing this, we can establish connections between new subtasks and existing subtasks (illustrated by dashed lines in Fig. 7.4), effectively creating Subtask nodes within the KG by identifying the most similar pairs. Subsequently, this paves the way for the prediction of relevant edges within the existing KG, serving as feasible configurations. These configurations supply context-based instructions to both human workers and robots involved in the HRC process.

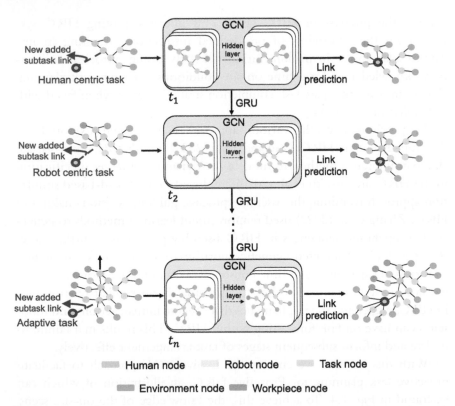

Figure 7.4 EvolveGCN method for proactive task planning in HRC. (Adapted from Zheng et al. (2022).)

In detail, the EvolveGCN approach combines GCN with Gated Recurrent Unit (GRU) for link prediction of dynamically changing human–robot roles within subtasks. During timestamp t, the GCN aggregation proceeds iteratively with the following equation:

$$H_t^{(l+1)} = \sigma(\hat{A}_t H_t^l W_t^l). \tag{7.1}$$

Here, $\hat{A}_t = \tilde{D}^{-\frac{1}{2}} \tilde{A}_t \tilde{D}^{\frac{1}{2}}$ represents the normalized version of an adjacency matrix A_t. The matrix $\tilde{A}_t = A_t + I$ is created by adding the identity matrix I to A_t. Additionally, $\tilde{D} = \text{diag}(\sum_j \tilde{A}_{t_{ij}})$ denotes the degree matrix, which is determined by the sum of elements in $\tilde{A}_{t_{ij}}$. The initial embedding matrix is derived from node features, while W_t^l signifies the weight matrix associated with layer l. This iterative process ensures the dynamic adaptation of the network to evolving human–robot roles and relationships in subtasks.

By analyzing the varying outcomes of link prediction within the KG, the system gains the ability to discern distinct HRC modes. These modes include human-centric, adaptive, or robot-centric approaches. To illustrate, when observing a new subtask at timestamp t_1, the system may identify it as a human-centric operation. This determination is based on the preponderance of linked nodes associated with human participants, meaning the dominating role of human operators. Similarly, when evaluating the situation at t_2, the system characterizes the operation as robot-centric. In this case, the nodes linked to robotic elements indicates the dominant involvement of the robot. Furthermore, at timestamp t_3, the system recognizes an adaptive operation, based on the diversity and types of linked nodes. This mode implies a flexible collaboration where the roles of humans and robots are dynamically adjusted based on contextual requirements. Consequently, by identifying the specific HRC mode, the Proactive HRC system not only provides instructions for the forthcoming stages to both humans and robots but also manages each party's permissions, avoiding potential conflicts such as the issuance of misleading orders by humans. This adaptive control is crucial for ensuring a fluent and productive collaboration between humans and robots in various operational scenarios.

Alternatively, utilizing the perceptual knowledge derived from the real-time HRC scene on-site, a temporal HRC KG can be effectively constructed. This dynamic KG facilitates the allocation of tasks to human operators and robots based on specific real-time requirements. The framework is known as the DynGraphGAN-GRU model (Xiong et al., 2020). By merging GCN, Generative Adversarial Network (GAN), and GRU, the model effectively captures both temporal and structural information of the data. This comprehensive understanding is then leveraged to predict the associations or links between different nodes within the KG. The DynGraphGAN-GRU model operates by continuously updating and evolving its knowledge representation based on real-time data. This dynamic and adaptive framework presented in Fig. 7.5 aids in determining task allocation between human operators and robots with a high degree of responsiveness, which is feasible in the ever-changing context of HRC scenarios. The real-time adjustments made by the model ensure that tasks are assigned to the most suitable entity, whether human or robot, for the ongoing collaboration, enhancing efficiency and safety in the HRC process.

Considering the diverse inputs covering human behaviors, object detections, and task structures, diverse possible configurations can be established to define their relationships. It is noted that even subtly different connections between the Human–Robot–Task–Workpiece–Environment (HRTWE) nodes can significantly impact the performance and applicability of these configurations. To adapt the HRC KG to the ever-changing dynamics of various scenarios over different timestamps, a continuous discriminator is incorporated into the DynGraphGAN-GRU model. The primary role of this discriminator is to differentiate between authentic and artificial edges within the HRTWE nodes. This distinction can update the KG to reflect the most current state of the HRC system. As the KG undergoes these updates, it facilitates the optimization of task arrangements and the establishment of sequential orders. A practical example of this process can be seen when the DynGraphGAN-GRU model detects signs of human fatigue due to repeated and high-load human operations. In response, it adjusts the KG by establishing connections between human nodes and specific subtask nodes that predominantly involve lighter workload operations. This adaptation ensures that the HRC system can efficiently re-distribute tasks, prioritize tasks more aligned with human capabilities, and maintain smooth operations in the face of changing human conditions.

In detailed DynGraphGAN-GRU algorithm, the inputs consist of H_t^l, a representation of node embeddings corresponding to a specific timestamp, t. These embeddings encapsulate both structural information and feature attributes of the nodes. Within this GAN framework, the Discriminator's primary function is to evaluate the input data and discern the authenticity of connections, which is essential for optimizing the learning process. The entire operation is guided by a loss function for the training

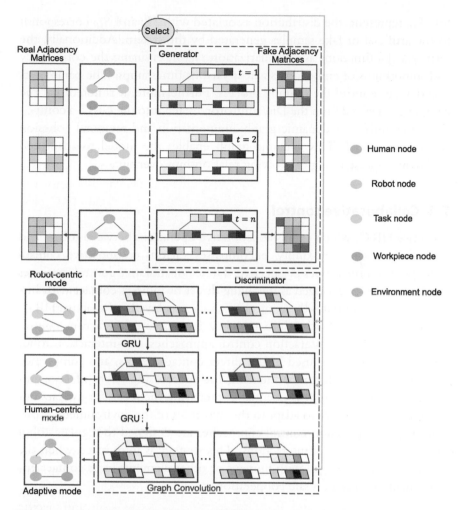

Figure 7.5 DynGraphGAN-GRU method for proactive task planning in Proactive HRC. (Adapted from Zheng et al. (2023).)

and convergence of the model. This loss function is formulated as follows:

$$\min_{G} \max_{D} V(D, G) = E_{v' \to p_v}[D(S_{H_t^l}) - D(G(H_t^l))]$$

$$+ \alpha E_{H_t^l \to P_{H_t^l}} \left(\left\| \nabla_{\hat{S}_{v'}} D(\hat{S}_{H_t^l}) \right\|_2 - 1 \right)^2 + \lambda \left\| \Delta_G \right\|_F^2. \tag{7.2}$$

Here, $G(H_t^l)$ represents a sequence of adjacency matrices generated by the generator, where both G and D denote the components of the Discriminator; $P_{H_t^l}$ represents the distribution associated with H_t^l; and $\hat{S}_{H_t^l}$ corresponds to the artificial or fake samples generated by the system. Additionally, the term $\|\Delta_G\|_F^2$ functions as a regularization factor, ensuring the consistency and smoothness of embeddings across various timestamps. The parameters α and λ can control the training process. Furthermore, it is noted that the parameters derived from the Discriminator are fed into the GRU, a component responsible for dynamic graph augmentation and temporal cohesion within the model. This dynamic graph enables the model to adapt and evolve its understanding of evolving relationships over time.

7.4 Collaborative control

Proactive HRC, which entails collaborative operations between robots and humans in a close proximity, encompasses various collaborative control strategies from the robot's perspective. These collaborative control strategies can be broadly categorized as involving direct or indirect contact with humans. Fig. 7.6 illustrates the control methods employed in Proactive HRC, which are further divided into two key aspects: pre-interaction control approaches and post-interaction control approaches. Pre-interaction robot control approaches are used before direct contact occurs, and compliance control is a pivotal factor influencing the success and safety of cooperative actions. This strategy is essential in preparing the robot for interactions and ensuring its ability to adapt to the dynamics of the environment. Conversely, post-interaction control approaches are designed to regulate robot motion during physical interactions. This strategy is introduced when the robot is actively engaged with humans or other elements in its contact environment, focusing on safe cooperation.

7.4.1 Pre-interaction robot control

Pre-interaction robot control involves using perceived and predicted human motions and environmental changes to adapt the robot's movements before direct interaction occurs. Various approaches, including orientation adaptation, speed regulation, trajectory modification, and the potential field method, have been applied in recent research to achieve the objective. Table 7.12 provides examples of these approaches. To address safety constraints in HRC, Zanchettin et al. (2015) proposed safety evaluation metrics based on the relative distance between the robot and the human operator. They introduced an optimization-based control strategy to modulate the robot's velocity along a designated path, ensuring both safety and efficiency in

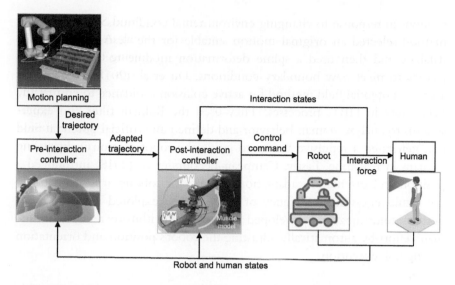

Figure 7.6 Collaborative control scheme for Proactive HRC.

Table 7.12 Typical research efforts on pre-interaction robot control.

Objective	Method	Reference
Speed regulation	Distance-based safety evaluation, and optimization-based control strategy to modulate robot velocity	Zanchettin et al. (2015)
Trajectory modification	Statically equivalent serial chain technique, and online trajectory optimization	Kim et al. (2017)
Trajectory modification	Spline decomposition, spline interpolation	Pekarovskiy et al. (2017)
Potential field method	Unscented Kalman filter, expert system, and artificial potential field	Du et al. (2018)
Position and orientation adaptation	Intuitive adaptive orientation control	Campeau-Lecours et al. (2018)

Proactive HRC. Kim et al. (2017) introduced an approach to adaptively control the robot's motion in real-time during human–robot load-sharing tasks to prevent human joint overloading. Their method estimated human joint load using a statically equivalent serial chain model and used online optimization techniques to modify the robot's motion, enabling ergonomic body positioning and reducing joint torque. Pekarovskiy et al. (2017) proposed a real-time spline deformation approach to generate feasible robot motions in response to changing environmental conditions or goals. Their method selected an original motion suitable for the desired task from a database and then used a spline deformation module to adjust the trajectory to meet new boundary conditions. Du et al. (2018) adopted the artificial potential field method for active collision avoidance to ensure human safety in HRC processes. They used the Kalman filter and expert systems to analyze human behavior and defined an artificial potential field that generated a repulsive vector to guide the robot motion in bypassing humans in real-time. Lastly, Campeau-Lecours et al. (2018) introduced a position and orientation adaptation scheme for robotic arms, enhancing the intuitiveness and efficiency of HRC. They explored different robot poses' intuitiveness and developed a control algorithm to reduce the human's effort by automatically adjusting the robot's position and orientation during task execution.

7.4.2 Post-interaction robot control

Post-interaction robot control involves the modulation of motion and contact force between humans and robots to enhance the collaboration experience. Popular approaches in this context are mainly based on admittance control and impedance control, as shown in the Table 7.13. Carmichael et al. (2017) explored an admittance control approach to ensure safe and comfortable HRI, especially in proximity to the robot's singular configurations. An exponentially shaped damping profile and an asymmetric damping scheme were introduced to ensure stability and smooth robot operation near singularities. Additionally, a repulsive force method was employed to guide the robot away from singular configurations. To address the instability problem that may occur when robots interact with rigid environments, Ferraguti et al. (2019) proposed a novel variable admittance control scheme. They detected increasing oscillations during Human–Robot Interaction (HRI) through statistical analysis and introduced a parameter adaptation strategy to restore stability while preserving the passivity of the admittance dynamics. Besides, to ensure that state and output constraints of the robot are enforced during HRI, Kimmel and Hirche (2017) developed an invariance control scheme for safe HRC in complex environments. They employed feedback linearization of human–robot dynamics to analytically determine a set of safe system states. An invariance control module was combined with an impedance control scheme to keep the system within the admissible set of states while bounding the tracking error compared to the initial trajectory. Then, Roveda et al. (2020) designed a model-based

Table 7.13 Typical research efforts on post-interaction robot control.

Objective	Method	Reference
Admittance control	Damped–least–squares with exponentially shaped damping profile, and repulsive force field method	Carmichael et al. (2017)
Admittance control	Instability detection, and passivity-based parameter adaptation	Ferraguti et al. (2019)
Impedance control	Invariance control, and feedback liberalization	Kimmel and Hirche (2017)
Impedance control	RL, model predictive control, and cross-entropy method	Roveda et al. (2020)
Impedance control	Impedance learning, parameters adaptation, and sliding mode control	Sharifi et al. (2021)

RL control scheme with variable impedance to account for the complex interaction dynamics in HRC. They captured the HRI dynamics with a neural network model and utilized a model predictive controller with the cross-entropy method to optimize impedance control parameters on-line, minimizing human effort. Lastly, Sharifi et al. (2021) introduced an impedance learning-based control approach to facilitate stable HRI without the direct measurement of interaction forces. They employed a learning policy to online adjust the robot's impedance based on human behavior. Two sets of updating laws were formulated to adjust the robot's dynamic parameters and controller gains, compensating for structured and unstructured uncertainties from both the robot and human.

7.4.3 An example of collaborative intelligence-based control

As articulated in Zheng et al. (2023), the Collaborative Intelligence (CI)-based approach offers a robust solution for addressing industrial uncertainties within Proactive HRC systems, while allowing for fine-tuned tolerance settings. CI achieves complementary between human intelligence and artificial intelligence through two key aspects:

a) (Human-Assisting-Robot) This aspect emphasizes humans taking the lead in training robots on how to adapt operations to new situations. Humans play a crucial role in explaining the decisions made by the robot and overseeing updates to task-related knowledge, particularly in the context of dealing with robot uncertainties.

b) (Robot-Assisting-Human) In this mode, robots step in to create a more conducive work environment for humans. They facilitate human interaction with tasks by offering flexibility in decision-making and considering human-centric requirements, especially when facing human uncertainties.

To enable humans to assist robots, El Zaatari et al. (2021) leveraged Learning from Demonstration (LfD). This approach allowed cobots to autonomously adjust to new environmental settings based on knowledge acquired from demonstrated paths. Going beyond control-level assistance, You et al. (2022) proposed task knowledge models extracted from dual-human demonstrations. This approach tracked and segmented video demonstrations into action sequences and structured them into a semantic model for handling assembly tasks effectively. On the flip side, to empower robots to consider human needs in HRC, Liu et al. (2021) integrated a parameterized reward function of safety considerations, into Deep Reinforcement Learning (DRL) models. This integration paved the way for robot motion planning that ensured safe human–robot co-working. Meanwhile, Quenehen et al. (2020) delved into hybrid collaboration modes between workers and robots, optimizing the assembly process while meeting ergonomic objectives. Despite these advancements, there remains a deficiency for efficient adaption of both human intelligence and acquired knowledge in the face of uncertainties in HRC.

Motivated by this challenge and with a focus on addressing HRC uncertainties, the proposed CI-based control strategy in Proactive HRC, as depicted in Fig. 7.7, provides a solution. When a human operator encounters task uncertainties or observes unexpected robot behaviors, the system allows for human-guided re-training and skill enhancement of the robot through flexible demonstrations using LfD techniques. Conversely, if the robot detects uncertain human activities during task execution, it can autonomously optimize its trajectories through a DRL-based approach to adapt to unforeseen human behaviors, ensuring safe and efficient collaboration. Consequently, the CI-based HRC approach maximizes human–robot complementarity and bolsters its resilience when confronted with uncertain task scenarios.

7.4.3.1 Human-assisted robot via LfD

Given the presence of robot and task uncertainties in the HRC domain, human operators possess the capability to transfer their expertise to refine robot manipulation skills, for flexible and adaptive task execution. In this context, the introduction of an LfD approach uses the intelligence of human experts to generate robot task control policies.

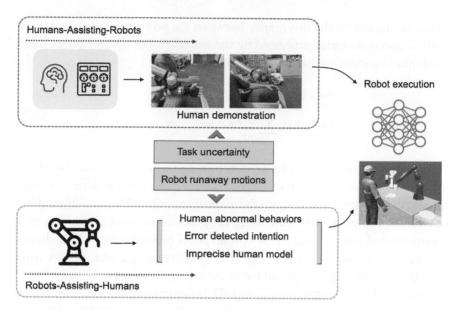

Figure 7.7 Collaborative intelligence-based control for handling uncertainties in Proactive HRC. (Adapted from Zheng et al. (2023).)

In programming robot control policies, there is a challenge when attempting to naturally convert the experiential knowledge of human experts into actionable robot control policies. To bridge this gap, the hand poses extracted from the HRC scene can be utilized to establish a seamless, hand-gesture-enabled robot control system. The worker's hand poses are extracted and then correlated with the robot's end-effector poses. This intuitive mapping enables the robot to mimic the human hand's movements.

Beyond explicit imitation, the movement trajectories demonstrated by expert human operators are recorded as datasets. These datasets are fed into LfD algorithms to implicitly extract the worker's patterns. The learned knowledge allows robots to adapt and develop control policies to handling uncertainties in subsequent automated task programming.

In the LfD approach, Behavioral Cloning (BC) algorithm is used to approximate the robot control policy $\pi_\theta(a|s)$. The dataset trajectories, denoted as $\tau_1, \tau_2, \ldots, \tau_m$, comprise environment observations s_1^i and robot action a_1^i. The state–action pair (s, a) encapsulates environmental context, task conditions, human-specific information, and robot actions. Each demonstration in the dataset can be represented as $\tau_i < \{s_1^i, a_i^i, s_2^i, a_2^i, \ldots, s_{n+1}^i\}$, and the dataset is presented as $D = \{(S_1, A_1), (S_2, A_2), (S_3, A_3), \ldots\}$. The BC algorithm can minimize the divergence between the probability distribution of state–action trajectories generated by the model (robot control policy) and the input trajectory probability distribution:

$$\max_{\theta} \mathbb{E}_{(s,a) \to D}[\pi_{\theta}(a|s)]$$

$$\text{such that } \sum_{a \in A} \hat{\pi}_{\theta}(a|s) = 1, \quad \forall s \in S. \tag{7.3}$$

Within the parameter optimization process of maximum likelihood estimation, a policy $\pi_{\theta}(s)$ is trained to minimize the discrepancies between observed behavioral patterns in the demonstrations. This is achieved by optimizing the policy function using deep neural networks, resulting in a parameterized policy $\pi_{\theta}(s)$. The optimization process relies on gradient-based methods to minimize the objective function $\mathbb{E}_{(s,a) \to D} \|\pi_{\theta}(a|s)\|^2$, with the goal of acquiring an optimal robot control policy function.

Nonetheless, the effectiveness of LfD is constrained by the diversity of the samples. It is influenced by both the number and variability of expert demonstrations. Meanwhile, the robot policy trained via the BC algorithm lacks the requisite flexibility and adaptability when confronted with new emerging uncertainties. To overcome this limitation, an online learning approach, complemented by the Dataset Aggregation (Dagger) mechanism, is introduced within the framework of LfD.

The Dagger mechanism can effectively address the new uncertainties that emerge during task execution and extend the learned policy to encompass new working scenarios. As a result, the robot can adeptly tackle manufacturing tasks in diverse work scenarios, handling HRC uncertainties.

7.4.3.2 Robot-assisted human via DRL

In response to human uncertainties like abnormal behaviors, the robot should dynamically adjust its motion plans in real-time to ensure both task completion and human safety. In this context, a DRL-based approach is introduced to achieve human uncertainty-aware robot control for safe and smooth HRC.

Various human uncertainty factors in HRC include whole-body skeleton positions, indications of abnormal behaviors, and discerned human intentions. In practice, the DRL approach leverages these uncertainty factors along with other parameters such as motion planning success rates and safety constraints within HRC scenarios as optimization metrics. For the robot motion planning process, a Markov Decision Process (MDP) is introduced. The objective is to optimize the control policy denoted as π^* and guide the robot's selection of actions, $a_t \in A$, temporal state $s_t \in S$, to maximize cumulative rewards. In the DRL settings, the following components are identified:

1. (Observation State, O) This represents the current human–robot working scene, including data related to whole-body skeleton positions (P), abnormal human behaviors (B), human intentions (I), and the robot's state (M). These parameters are concatenated into a state vector, denoted as $S = (P, B, I, M)$.

2. (Action Space, A) It defines the range of actions accessible to the robot. Specifically, it relates to the robot's reachability, and involves utilizing inverse kinematics to transform the robot's joint space into three-dimensional spatial coordinates of the end effector, represented as $A = (X, Y, Z) \in \mathbb{R}^3$.

3. (Reward, R) This component includes multiple evaluation criteria, which comprise safety considerations (such as distance and collision states) and evaluations of task completion progress, including success rates. These are represented as $R = (R_s, R_t)$.

In implementation, the actor–critic framework is employed to facilitate learning and control actions. This approach can maximize the expected return $J(\theta)$ and optimize the robot's path by emphasizing safety concerns. Here, $p_\theta(\tau) = p(s_0) \prod_{t=0}^{T-1} [(s_{t+1}|s_t, a_t)\pi_\theta(a_t|s_t)]$ denotes the probability distribution across all potential state–action trajectories $\tau = (s_0, a_0, s_1, \ldots, a_{T-1}, s_T)$. Additionally, $\gamma^t \in [0, 1]$ represents the discount factor at time t, and $d_\theta(s_t)$ stands for the state distribution under the policy θ_π:

$$J(\theta) = \mathbb{E}_{\tau \, > p_\theta(t)} [\sum_{t=0}^{T} \gamma^t r_t]. \tag{7.4}$$

7.5 Chapter summary

This chapter illustrates the development of a Proactive HRC system, which integrates four modules, including scene perception, knowledge representation learning, decision-making, and collaborative control. These modules collectively contribute to the establishment of an efficient and effective workflow for Proactive HRC in the field of smart manufacturing. The algorithms presented within these modules serve as valuable tools for researchers and engineers in the smart manufacturing domain, offering insights into the implementation of Proactive HRC. Building upon this foundational knowledge, participants are encouraged to delve into the distinctive attributes of Proactive HRC, aligning with the human-centric requirements of production. These attributes encompass mutual-cognitive capabilities, predictability, and self-organizing features, as demonstrated in preceding chapters.

References

Abdelkawy, H., Ayari, N., Chibani, A., Amirat, Y., Attal, F., 2020. Spatio-temporal convolutional networks and n-ary ontologies for human activity-aware robotic system. IEEE Robotics and Automation Letters 6, 620–627.

Aggravi, M., Elsherif, A.A.S., Giordano, P.R., Pacchierotti, C., 2021. Haptic-enabled decentralized control of a heterogeneous human–robot team for search and rescue in partially-known environments. IEEE Robotics and Automation Letters 6, 4843–4850.

Ahn, H., Choi, S., Kim, N., Cha, G., Oh, S., 2018. Interactive text2pickup networks for natural language-based human–robot collaboration. IEEE Robotics and Automation Letters 3, 3308–3315.

Aldini, S., Akella, A., Singh, A.K., Wang, Y.K., Carmichael, M., Liu, D., Lin, C.T., 2019. Effect of mechanical resistance on cognitive conflict in physical human–robot collaboration. In: 2019 International Conference on Robotics and Automation (ICRA). IEEE, pp. 6137–6143.

Ansari, R.J., Karayiannidis, Y., 2021. Task-based role adaptation for human–robot cooperative object handling. IEEE Robotics and Automation Letters 6, 3592–3598.

Arkin, J., Park, D., Roy, S., Walter, M.R., Roy, N., Howard, T.M., Paul, R., 2020. Multimodal estimation and communication of latent semantic knowledge for robust execution of robot instructions. The International Journal of Robotics Research 39, 1279–1304.

Arkouli, Z., Kokotinis, G., Michalos, G., Dimitropoulos, N., Makris, S., 2021. AI-enhanced cooperating robots for reconfigurable manufacturing of large parts. IFAC-PapersOnLine 54, 617–622.

Birch, B., Griffiths, C., Morgan, A., 2021. Environmental effects on reliability and accuracy of MFCC based voice recognition for industrial human–robot interaction. Proceedings of the Institution of Mechanical Engineers. Part B, Journal of Engineering Manufacture 235, 1939–1948.

Buerkle, A., Bamber, T., Lohse, N., Ferreira, P., 2021a. Feasibility of detecting potential emergencies in symbiotic human–robot collaboration with a mobile EEG. Robotics and Computer-Integrated Manufacturing 72, 102179.

Buerkle, A., Eaton, W., Lohse, N., Bamber, T., Ferreira, P., 2021b. EEG based arm movement intention recognition towards enhanced safety in symbiotic human–robot collaboration. Robotics and Computer-Integrated Manufacturing 70, 102137.

Buerkle, A., Matharu, H., Al-Yacoub, A., Lohse, N., Bamber, T., Ferreira, P., 2022. An adaptive human sensor framework for human–robot collaboration. The International Journal of Advanced Manufacturing Technology 119, 1233–1248.

Campeau-Lecours, A., Côté-Allard, U., Vu, D.S., Routhier, F., Gosselin, B., Gosselin, C., 2018. Intuitive adaptive orientation control for enhanced human–robot interaction. IEEE Transactions on Robotics 35, 509–520.

Carmichael, M.G., Liu, D., Waldron, K.J., 2017. A framework for singularity-robust manipulator control during physical human–robot interaction. The International Journal of Robotics Research 36, 861–876.

Casalino, A., Messeri, C., Pozzi, M., Zanchettin, A.M., Rocco, P., Prattichizzo, D., 2018. Operator awareness in human–robot collaboration through wearable vibrotactile feedback. IEEE Robotics and Automation Letters 3, 4289–4296.

Chen, C.S., Chen, S.K., Lai, C.C., Lin, C.T., 2020. Sequential motion primitives recognition of robotic arm task via human demonstration using hierarchical BiLSTM classifier. IEEE Robotics and Automation Letters 6, 502–509.

Cheng, Q., Zhang, W., Liu, H., Zhang, Y., Hao, L., 2021. Research on the path planning algorithm of a manipulator based on GMM/GMR-MPRM. Applied Sciences 11, 7599.

Cheng, Y., Sun, L., Liu, C., Tomizuka, M., 2020. Towards efficient human–robot collaboration with robust plan recognition and trajectory prediction. IEEE Robotics and Automation Letters 5, 2602–2609.

Dalmasso, M., Garrell, A., Domínguez, J.E., Jiménez, P., Sanfeliu, A., 2021. Human–robot collaborative multi-agent path planning using Monte Carlo tree search and social reward sources. In: 2021 IEEE International Conference on Robotics and Automation (ICRA). IEEE, pp. 10133–10138.

Dias, A., Wellaboda, H., Rasanka, Y., Munasinghe, M., Rodrigo, R., Jayasekara, P., 2020. Deep learning of augmented reality based human interactions for automating a robot team. In: 2020 6th International Conference on Control, Automation and Robotics (ICCAR). IEEE, pp. 175–182.

Ding, Y., Xu, W., Liu, Z., Zhou, Z., Pham, D.T., 2019. Robotic task oriented knowledge graph for human–robot collaboration in disassembly. Procedia CIRP 83, 105–110.

Du, G., Long, S., Li, F., Huang, X., 2018. Active collision avoidance for human–robot interaction with UKF, expert system, and artificial potential field method. Frontiers in Robotics and AI 5, 125.

El Makrini, I., Mathijssen, G., Verhaegen, S., Verstraten, T., Vanderborght, B., 2022. A virtual element-based postural optimization method for improved ergonomics during human–robot collaboration. IEEE Transactions on Automation Science and Engineering.

El Zaatari, S., Wang, Y., Li, W., Peng, Y., 2021. ITP-LFD: Improved task parametrised learning from demonstration for adaptive path generation of cobot. Robotics and Computer-Integrated Manufacturing 69, 102109.

Fan, J., Zheng, P., Lee, C.K., 2022a. A multi-granularity scene segmentation network for human–robot collaboration environment perception. In: 2022 IEEE/RSJ International Conference on Intelligent Robots and Systems (IROS). IEEE, pp. 2105–2110.

Fan, J., Zheng, P., Lee, C.K., 2023. A vision-based human digital twin modelling approach for adaptive human–robot collaboration. Journal of Manufacturing Science and Engineering, 1–11.

Fan, J., Zheng, P., Li, S., Wang, L., 2022b. An integrated hand–object dense pose estimation approach with explicit occlusion awareness for human–robot collaborative disassembly. IEEE Transactions on Automation Science and Engineering.

Ferraguti, F., Talignani Landi, C., Sabattini, L., Bonfè, M., Fantuzzi, C., Secchi, C., 2019. A variable admittance control strategy for stable physical human–robot interaction. The International Journal of Robotics Research 38, 747–765.

Franceschi, P., Castaman, N., Ghidoni, S., Pedrocchi, N., 2020. Precise robotic manipulation of bulky components. IEEE Access 8, 222476–222485.

Gielniak, M.J., Thomaz, A.L., 2011. Generating anticipation in robot motion. In: 2011 RO-MAN. IEEE, pp. 449–454.

Gori, I., Aggarwal, J., Matthies, L., Ryoo, M.S., 2016. Multitype activity recognition in robot-centric scenarios. IEEE Robotics and Automation Letters 1, 593–600.

Granados, D.F.P., Yamamoto, B.A., Kamide, H., Kinugawa, J., Kosuge, K., 2017. Dance teaching by a robot: Combining cognitive and physical human–robot interaction for supporting the skill learning process. IEEE Robotics and Automation Letters 2, 1452–1459.

Guo, Y., Yang, X.J., 2021. Modeling and predicting trust dynamics in human–robot teaming: A Bayesian inference approach. International Journal of Social Robotics 13, 1899–1909.

He, K., Zhang, X., Ren, S., Sun, J., 2016. Deep residual learning for image recognition. In: Proceedings of the IEEE Conference on Computer Vision and Pattern Recognition, pp. 770–778.

He, Y., Li, F., Li, J., Liu, J., Wu, X., 2022. An SEMG based adaptive method for human-exoskeleton collaboration in variable walking environments. Biomedical Signal Processing and Control 74, 103477.

Hetherington, N.J., Croft, E.A., Van der Loos, H.M., 2021. Hey robot, which way are you going? Nonverbal motion legibility cues for human–robot spatial interaction. IEEE Robotics and Automation Letters 6, 5010–5015.

Hietanen, A., Pieters, R., Lanz, M., Latokartano, J., Kämäräinen, J.K., 2020. AR-based interaction for human–robot collaborative manufacturing. Robotics and Computer-Integrated Manufacturing 63, 101891.

Hignett, S., McAtamney, L., 2000. Rapid entire body assessment (REBA). Applied Ergonomics 31, 201–205.

Hoang, V.T., Jo, K.H., 2018. 3-D human pose estimation using cascade of multiple neural networks. IEEE Transactions on Industrial Informatics 15, 2064–2072.

Hu, Z., Pan, J., Fan, T., Yang, R., Manocha, D., 2019. Safe navigation with human instructions in complex scenes. IEEE Robotics and Automation Letters 4, 753–760.

Karami, H., Darvish, K., Mastrogiovanni, F., 2020. A task allocation approach for human-robot collaboration in product defects inspection scenarios. In: 2020 29th IEEE International Conference on Robot and Human Interactive Communication (RO-MAN). IEEE, pp. 1127–1134.

Kato, H., Ushiku, Y., Harada, T., 2018. Neural 3D mesh renderer. In: Proceedings of the IEEE Conference on Computer Vision and Pattern Recognition, pp. 3907–3916.

Khatib, M., Al Khudir, K., De Luca, A., 2021. Human–robot contactless collaboration with mixed reality interface. Robotics and Computer-Integrated Manufacturing 67, 102030.

Kim, W., Lee, J., Peternel, L., Tsagarakis, N., Ajoudani, A., 2017. Anticipatory robot assistance for the prevention of human static joint overloading in human–robot collaboration. IEEE Robotics and Automation Letters 3, 68–75.

Kimmel, M., Hirche, S., 2017. Invariance control for safe human–robot interaction in dynamic environments. IEEE Transactions on Robotics 33, 1327–1342.

Lanini, J., Razavi, H., Urain, J., Ijspeert, A., 2018. Human intention detection as a multiclass classification problem: Application in physical human–robot interaction while walking. IEEE Robotics and Automation Letters 3, 4171–4178.

Lee, H., Liau, Y., Kim, S., Ryu, K., 2018. A framework for process model based human-robot collaboration system using augmented reality. In: IFIP International Conference on Advances in Production Management Systems. Springer, pp. 482–489.

Lee, M.L., Behdad, S., Liang, X., Zheng, M., 2020. Disassembly sequence planning considering human–robot collaboration. In: 2020 American Control Conference (ACC). IEEE, pp. 2438–2443.

Liau, Y.Y., Ryu, K., 2022. Genetic algorithm-based task allocation in multiple modes of human–robot collaboration systems with two cobots. The International Journal of Advanced Manufacturing Technology 119, 7291–7309.

Lin, T.Y., Dollár, P., Girshick, R., He, K., Hariharan, B., Belongie, S., 2017. Feature pyramid networks for object detection. In: Proceedings of the IEEE Conference on Computer Vision and Pattern Recognition, pp. 2117–2125.

Lin, W., Anwar, A., Li, Z., Tong, M., Qiu, J., Gao, H., 2018. Recognition and pose estimation of auto parts for an autonomous spray painting robot. IEEE Transactions on Industrial Informatics 15, 1709–1719.

Liu, H., Wang, L., 2021. Collision-free human–robot collaboration based on context aware-ness. Robotics and Computer-Integrated Manufacturing 67, 101997.

Liu, Q., Liu, Z., Xiong, B., Xu, W., Liu, Y., 2021. Deep reinforcement learning-based safe interaction for industrial human–robot collaboration using intrinsic reward function. Advanced Engineering Informatics 49, 101360.

Liu, S., Wang, L., Vincent Wang, X., 2022a. Multimodal data-driven robot control for human–robot collaborative assembly. Journal of Manufacturing Science and Engineering 144, 051012.

Liu, X., Zheng, L., Shuai, J., Zhang, R., Li, Y., 2020. Data-driven and AR assisted intelli-gent collaborative assembly system for large-scale complex products. Procedia CIRP 93, 1049–1054.

Liu, Y., Habibnezhad, M., Jebelli, H., 2022b. Worker-aware task planning for construc-tion robots: A physiologically based communication channel interface. In: Automation and Robotics in the Architecture, Engineering, and Construction Industry. Springer, pp. 181–200.

Liu, Z., Mao, H., Wu, C.Y., Feichtenhofer, C., Darrell, T., Xie, S., 2022c. A ConvNet for the 2020s. In: Proceedings of the IEEE/CVF Conference on Computer Vision and Pattern Recognition, pp. 11976–11986.

Loper, M., Mahmood, N., Romero, J., Pons-Moll, G., Black, M.J., 2023. SMPL: A skinned multi-person linear model. In: Seminal Graphics Papers: Pushing the Boundaries, vol. 2, pp. 851–866.

Lorenzini, M., Kim, W., De Momi, E., Ajoudani, A., 2019. A new overloading fa-tigue model for ergonomic risk assessment with application to human–robot collabo-ration. In: 2019 International Conference on Robotics and Automation (ICRA). IEEE, pp. 1962–1968.

Losey, D.P., O'Malley, M.K., 2019. Enabling robots to infer how end-users teach and learn through human–robot interaction. IEEE Robotics and Automation Letters 4, 1956–1963.

Lu, L., Wang, H., Reily, B., Zhang, H., 2021. Robust real-time group activity recognition of robot teams. IEEE Robotics and Automation Letters 6, 2052–2059.

Macchini, M., Schiano, F., Floreano, D., 2019. Personalized telerobotics by fast ma-chine learning of body–machine interfaces. IEEE Robotics and Automation Letters 5, 179–186.

Maderna, R., Pozzi, M., Zanchettin, A.M., Rocco, P., Prattichizzo, D., 2022. Flexible scheduling and tactile communication for human–robot collaboration. Robotics and Computer-Integrated Manufacturing 73, 102233.

Malik, A.A., Brem, A., 2021. Digital twins for collaborative robots: A case study in human–robot interaction. Robotics and Computer-Integrated Manufacturing 68, 102092.

Marín, R., Sanz, P.J., Nebot, P., Wirz, R., 2005. A multimodal interface to control a robot arm via the web: A case study on remote programming. IEEE Transactions on Industrial Electronics 52, 1506–1520.

Marullo, S., Pozzi, M., Prattichizzo, D., Malvezzi, M., 2020. Cooperative human–robot grasping with extended contact patches. IEEE Robotics and Automation Letters 5, 3121–3128.

Mateus, J.C., Claeys, D., Limère, V., Cottyn, J., Aghezzaf, E.H., 2020. Base part centered assembly task precedence generation. The International Journal of Advanced Manufac-turing Technology 107, 607–616.

Mateus, J.E.C., Claeys, D., Limère, V., Cottyn, J., Aghezzaf, E.H., 2019. Ergonomic and performance factors for human–robot collaborative workplace design and evaluation. IFAC-PapersOnLine 52, 2550–2555.

Moon, J., Lee, B., 2018. Scene understanding using natural language description based on 3D semantic graph map. Intelligent Service Robotics 11, 347–354.

Nikolaidis, S., Hsu, D., Srinivasa, S., 2017. Human–robot mutual adaptation in collaborative tasks: Models and experiments. The International Journal of Robotics Research 36, 618–634.

Oyekan, J.O., Hutabarat, W., Tiwari, A., Grech, R., Aung, M.H., Mariani, M.P., López-Dávalos, L., Ricaud, T., Singh, S., Dupuis, C., 2019. The effectiveness of virtual environments in developing collaborative strategies between industrial robots and humans. Robotics and Computer-Integrated Manufacturing 55, 41–54.

Parsa, B., Samani, E.U., Hendrix, R., Devine, C., Singh, S.M., Devasia, S., Banerjee, A.G., 2019. Toward ergonomic risk prediction via segmentation of indoor object manipulation actions using spatiotemporal convolutional networks. IEEE Robotics and Automation Letters 4, 3153–3160.

Pekarovskiy, A., Nierhoff, T., Hirche, S., Buss, M., 2017. Dynamically consistent online adaptation of fast motions for robotic manipulators. IEEE Transactions on Robotics 34, 166–182.

Peternel, L., Fang, C., Tsagarakis, N., Ajoudani, A., 2019. A selective muscle fatigue management approach to ergonomic human–robot co-manipulation. Robotics and Computer-Integrated Manufacturing 58, 69–79.

Peternel, L., Tsagarakis, N., Caldwell, D., Ajoudani, A., 2018. Robot adaptation to human physical fatigue in human–robot co-manipulation. Autonomous Robots 42, 1011–1021.

Pulikottil, T.B., Pellegrinelli, S., Pedrocchi, N., 2021. A software tool for human–robot shared-workspace collaboration with task precedence constraints. Robotics and Computer-Integrated Manufacturing 67, 102051.

Quenehen, A., Thiery, S., Klement, N., Roucoules, L., Gibaru, O., 2020. Assembly process design: Performance evaluation under ergonomics consideration using several robot collaboration modes. In: Advances in Production Management Systems. Towards Smart and Digital Manufacturing: IFIP WG 5.7 International Conference, APMS 2020, Novi Sad, Serbia, August 30–September 3, 2020, Proceedings, Part II. Springer, pp. 477–484.

Raatz, A., Blankemeyer, S., Recker, T., Pischke, D., Nyhuis, P., 2020. Task scheduling method for HRC workplaces based on capabilities and execution time assumptions for robots. CIRP Annals 69, 13–16.

Rahman, S.M., Liao, Z., Jiang, L., Wang, Y., 2016. A regret-based autonomy allocation scheme for human–robot shared vision systems in collaborative assembly in manufacturing. In: 2016 IEEE International Conference on Automation Science and Engineering (CASE). IEEE, pp. 897–902.

Redmon, J., Farhadi, A., 2018. YOLOv3: An incremental improvement. arXiv:1804.02767.

Ren, S., He, K., Girshick, R., Sun, J., 2015. Faster R-CNN: Towards real-time object detection with region proposal networks. Advances in Neural Information Processing Systems 28.

Rey, R., Cobano, J.A., Corzetto, M., Merino, L., Alvito, P., Caballero, F., 2021. A novel robot co-worker system for paint factories without the need of existing robotic infrastructure. Robotics and Computer-Integrated Manufacturing 70, 102122.

Riaz, H., Terra, A., Raizer, K., Inam, R., Hata, A., 2020. Scene understanding for safety analysis in human–robot collaborative operations. In: 2020 6th International Conference on Control, Automation and Robotics (ICCAR). IEEE, pp. 722–731.

Riley, H., Sridharan, M., 2019. Integrating non-monotonic logical reasoning and inductive learning with deep learning for explainable visual question answering. Frontiers in Robotics and AI 6, 125.

Romero, J., Tzionas, D., Black, M.J., 2022. Embodied hands: Modeling and capturing hands and bodies together. arXiv:2201.02610.

Rosenberger, P., Cosgun, A., Newbury, R., Kwan, J., Ortenzi, V., Corke, P., Grafinger, M., 2020. Object-independent human-to-robot handovers using real time robotic vision. IEEE Robotics and Automation Letters 6, 17–23.

Roveda, L., Maskani, J., Franceschi, P., Abdi, A., Braghin, F., Tosatti, L.M., Pedrocchi, N., 2020. Model-based reinforcement learning variable impedance control for human–robot collaboration. Journal of Intelligent & Robotic Systems 100, 417–433.

Rozenberszki, D., Sörös, G., Szeier, S., Lőrincz, A., 2021. 3D semantic label transfer in human–robot collaboration. In: Proceedings of the IEEE/CVF International Conference on Computer Vision, pp. 2602–2611.

Sasagawa, A., Fujimoto, K., Sakaino, S., Tsuji, T., 2020. Imitation learning based on bilateral control for human–robot cooperation. IEEE Robotics and Automation Letters 5, 6169–6176.

Sauer, V., Sauer, A., Mertens, A., 2021. Zoomorphic gestures for communicating cobot states. IEEE Robotics and Automation Letters 6, 2179–2185.

Seichter, D., Köhler, M., Lewandowski, B., Wengefeld, T., Gross, H.M., 2021. Efficient RGB-D semantic segmentation for indoor scene analysis. In: 2021 IEEE International Conference on Robotics and Automation (ICRA). IEEE, pp. 13525–13531.

Sharifi, M., Azimi, V., Mushahwar, V.K., Tavakoli, M., 2021. Impedance learning-based adaptive control for human–robot interaction. IEEE Transactions on Control Systems Technology.

Simao, M.A., Gibaru, O., Neto, P., 2019. Online recognition of incomplete gesture data to interface collaborative robots. IEEE Transactions on Industrial Electronics 66, 9372–9382.

Sun, Y., Wang, W., Chen, Y., Jia, Y., 2020. Learn how to assist humans through human teaching and robot learning in human–robot collaborative assembly. IEEE Transactions on Systems, Man and Cybernetics: Systems.

Tan, H.L., Leong, M.C., Xu, Q., Li, L., Fang, F., Cheng, Y., Gauthier, N., Sun, Y., Lim, J.H., 2020. Task-oriented multi-modal question answering for collaborative applications. In: 2020 IEEE International Conference on Image Processing (ICIP). IEEE, pp. 1426–1430.

Tang, G., Webb, P., Thrower, J., 2019. The development and evaluation of robot light skin: A novel robot signalling system to improve communication in industrial human–robot collaboration. Robotics and Computer-Integrated Manufacturing 56, 85–94.

Tonola, C., Faroni, M., Pedrocchi, N., Beschi, M., 2021. Anytime informed path replanning and optimization for human–robot collaboration. In: 2021 30th IEEE International Conference on Robot & Human Interactive Communication (RO-MAN). IEEE, pp. 997–1002.

Tsarouchi, P., Matthaiakis, S.A., Michalos, G., Makris, S., Chryssolouris, G., 2016. A method for detection of randomly placed objects for robotic handling. CIRP Journal of Manufacturing Science and Technology 14, 20–27.

Tuli, T.B., Kohl, L., Chala, S.A., Manns, M., Ansari, F., 2021. Knowledge-based digital twin for predicting interactions in human–robot collaboration. In: 2021 26th IEEE International Conference on Emerging Technologies and Factory Automation (ETFA). IEEE, pp. 1–8.

Wang, L., Liu, S., Cooper, C., Wang, X.V., Gao, R.X., 2021. Function block-based human–robot collaborative assembly driven by brainwaves. CIRP Annals 70, 5–8.

Wang, P., Liu, H., Wang, L., Gao, R.X., 2018a. Deep learning-based human motion recognition for predictive context-aware human–robot collaboration. CIRP Annals 67, 17–20.

Wang, Q., Jiao, W., Yu, R., Johnson, M.T., Zhang, Y., 2019. Modeling of human welders' operations in virtual reality human–robot interaction. IEEE Robotics and Automation Letters 4, 2958–2964.

Wang, W., Li, R., Chen, Y., Diekel, Z.M., Jia, Y., 2018b. Facilitating human–robot collaborative tasks by teaching–learning collaboration from human demonstrations. IEEE Transactions on Automation Science and Engineering 16, 640–653.

Wang, X., Ong, S., Nee, A.Y.C., 2016. Multi-modal augmented-reality assembly guidance based on bare-hand interface. Advanced Engineering Informatics 30, 406–421.

Wang, X.V., Wang, L., Lei, M., Zhao, Y., 2020. Closed-loop augmented reality towards accurate human–robot collaboration. CIRP Annals 69, 425–428.

Xiong, Q., Zhang, J., Wang, P., Liu, D., Gao, R.X., 2020. Transferable two-stream convolutional neural network for human action recognition. Journal of Manufacturing Systems.

You, Y., Ji, Z., Yang, X., Liu, Y., 2022. From human–human collaboration to human–robot collaboration: Automated generation of assembly task knowledge model. In: 2022 27th International Conference on Automation and Computing (ICAC). IEEE, pp. 1–6.

Yu, T., Huang, J., Chang, Q., 2020. Mastering the working sequence in human–robot collaborative assembly based on reinforcement learning. IEEE Access 8, 163868–163877.

Zanchettin, A.M., Ceriani, N.M., Rocco, P., Ding, H., Matthias, B., 2015. Safety in human–robot collaborative manufacturing environments: Metrics and control. IEEE Transactions on Automation Science and Engineering 13, 882–893.

Zhang, R., Lv, Q., Li, J., Bao, J., Liu, T., Liu, S., 2022. A reinforcement learning method for human–robot collaboration in assembly tasks. Robotics and Computer-Integrated Manufacturing 73, 102227.

Zheng, P., Li, S., Fan, J., Li, C., Wang, L., 2023. A collaborative intelligence-based approach for handling human–robot collaboration uncertainties. CIRP Annals.

Zheng, P., Li, S., Xia, L., Wang, L., Nassehi, A., 2022. A visual reasoning-based approach for mutual-cognitive human–robot collaboration. CIRP Annals.

CHAPTER 8

Operational modes of industrial proactive human–robot collaboration

Jianzhuang Zhao[a,b] and Edoardo Lamon[a,c]

[a]Human–Robot Interfaces and Interaction, Italian Institute of Technology, Genoa, Italy
[b]Department of Electronics, Information and Bioengineering, Politecnico di Milano, Milan, Italy
[c]Department of Information Engineering and Computer Science, University of Trento, Trento, Italy

One of the primary impediments to the adoption of collaborative applications within industrial settings is closely associated with the current, albeit transitory, limitations in the development of AI systems responsible for regulating the behavior of robots. These limitations pertain to the aspects of robustness, flexibility, and comprehensibility in the solutions provided by AI. Specifically, these constraints hinder the ability of collaborative robots to determine their precise roles, particularly within complex tasks characterized by extended sequences of actions. Additionally, discerning the most advantageous coupling of tasks and agents (robotic or human worker) remains a challenge. Especially in small- and medium-sized enterprises (SMEs), the determination of whether to program a robot or delegate a task to a human often hinges on economic considerations linked to the production batch size. This economic-driven decision-making process underscores the requirement for the implementation of allocation strategies for multi-agent systems capable of designating the most suitable agent for the execution of each component of a task.

The underlying hypothesis in task allocation posits that the robot is capable of achieving each task independently, with human interactions limited to cooperative endeavors. However, there are situations where complete autonomous task execution by the robot may not be feasible. This limitation can stem from various factors, such as constraints within the robot's perception systems. In such instances, the human collaborator might need to engage in physical interactions with the robot to bridge the gap between the robot's autonomous capabilities and the task requirements. Thus, the development of suitable human–robot interfaces and physical interac-

Proactive Human–Robot Collaboration Toward Human-Centric Smart Manufacturing
https://doi.org/10.1016/B978-0-44-313943-7.00015-6

tive skills transfer methodologies is paramount for a fruitful human–robot collaboration.

In this chapter, we will demonstrate the typical operation modes of Proactive HRC in industrial scenarios, such as collaborative hierarchical/sequential operations using task allocation, human–robot comanipulation, and physical interactive tasks. The methods and results of this chapter are published in (Lamon et al., 2019; Merlo et al., 2023; Gandarias et al., 2022; Zhao et al., 2022).

8.1 Collaborative hierarchical and sequential operations

8.1.1 Connotation

In light of the advantages offered by collaborative solutions in workcell settings, researchers have made concerted efforts to bridge the gap between the stringent requirements of continuously hierarchical, sequential operations industrial tasks (e.g., assembly) and the paradigm of HRC. Their focus has primarily centered on enhancing human–robot interfaces, refining control modalities, and augmenting the robot's ability to perceive and understand the psycho-physical state of human collaborators. These endeavors aim to formulate effective collaboration plans that align with industrial needs.

A pivotal challenge in formulating a strategy for teamwork lies in the allocation of roles among the members of the collaborative team. This challenge is frequently referred to in the literature as role allocation, a problem predominantly explored within the domain of multirobot systems, where the harmonization of team efforts is essential to achieve a common overarching objective (Zhang et al., 2023). The customary approach involves the development of functions, often denoted as "utilities," that can assess the quality of matching agents with specific tasks and subsequently identify the optimal solution using optimization algorithms (Fusaro et al., 2021a; Merlo et al., 2022; El Makrini et al., 2022). However, the calculation of an agent's utility is not a straightforward task. It hinges on several factors, including the agent's competency to fulfill the responsibilities associated with a given role, the priority attributed to that role, and the dynamic needs of the team, both in the present moment and in future scenarios.

Extending the principles of role allocation from a multirobot setting to a heterogeneous team comprising both robots and human workers necessitates the formulation of a utility function for human workers. This utility function must be designed in a manner that allows it to be consistent with and comparable to the utility functions used to evaluate the suitability of robots for specific roles within the team. Another challenge relies on the choice of the design, which models all the possible collaborations among agents in HRC scenarios.

8.1.2 Role allocation strategy

A prevalent method for representing complex tasks and breaking them down into a series of fundamental actions within industrial assembly processes makes use of AND/OR graphs. This particular framework serves as a resourceful implementation of a state transition graph, boasting advantages such as a reduced node count and streamlined facilitation of the exploration of viable task plans. Moreover, the hyperarc weights enable optimal plan computation by means of graph-based inspection algorithms.

Hierarchical Assembly Modeling. Given a hierarchical assembly task P made of M pieces ($P = \{p_1, p_2, \ldots, p_M\}$), a configuration Θ of P is a set of subassemblies of P such that:

- **(i)**   each subassembly in Θ is formed by physically feasible and stable connections;
- **(ii)** each p_i belongs to one subassembly of Θ;
- **(iii)** a p_i cannot belong to two or more subassemblies of Θ.

The hierarchical assembly plan can be seen as a sequence of Θs, starting from $\Theta_i = \{\{p_1\}, \{p_2\}, \ldots, \{p_M\}\}$ and ending with $\Theta_f = \{\{p_1, p_2, \ldots, p_M\}\}$, i.e., starting from the configuration in which all the atomic pieces are separated (each of them is a subassembly), and ending with a unique final subset, corresponding to a configuration with all the pieces assembled. By considering feasible subassemblies as nodes and assembly operations that describe the transition between two configurations as edges, it is possible to find a path from Θ_i to Θ_f using path search algorithms. The edges, called *hyperarcs*, are pairs in which the first element is a single node (the father) and the second element is a set of nodes (the children); children represent all the possible subassemblies that can be obtained by disassembling the father node. For hierarchical assembly tasks, since most of the assembly operations consist of joining two subassemblies, hyperarcs are modeled as two-to-one connectors. Thus, an AOG is described by a set of nodes $N = \{n_1, n_2, \ldots, n_{|N|}\}$, and a set of hyperarcs $H = \{h_1, h_2, \ldots, h_{|H|}\}$. Each node $n \in N$ represents a subassembly of P, while hyperarcs define the assembly operations and can be characterized by different costs (Chang and Slagle, 1971). Children connected by the same h are in a logical AND, while

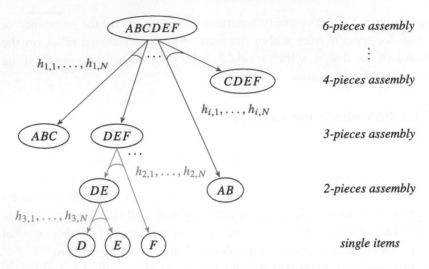

Figure 8.1 General AOG of an assembly made up of six pieces: A, B, C, D, E, and F. Each hyperarc has a different color since it represents a different sequential assembly action. Each hyperarc is duplicated to represent the feasibility of the action by each agent. Each arc has a different cost according to the suitability of each agent for performing the corresponding action.

different hyperarcs with the same parent node are in a logical OR. In general, it is possible to obtain a specific assembly configuration $\bar{\Theta}$ performing assembly operations between different subassemblies. Such subassemblies are in the AND relation, while the different assembly operations are represented by hyperarcs in the OR relation. For example, as described in Fig. 8.1, to complete the assembly $ABCDEF$, it is possible to join ABC with DEF (assembly operation described by hyperarc h_1) or join AB with $CDEF$ represented by h_i. Thus, ABC is linked in AND with DEF and AB with $CDEF$, while h_1 and h_i are in a logical OR. The only node without a father is named root. The nodes without children are identified as leaf nodes, and they are as many as the assembly pieces.

Including Human–Robot Collaboration. The standard AOG definition was augmented with two supplementary sets: the set of workers involved in the task execution, $W = \{w_1, w_2, \ldots, w_{|W|}\}$, and the set of actions that have to be performed, $A = \{a_1, a_2, \ldots, a_{|A|}\}$. The assembly sequence allocated to each worker is obtained by duplicating the same hyperarc (that describes an assembly operation) for $|W|$ times and assigning a cost to each $h_i \in H$, i.e., c_{h_i, w_j} with $i \in [1, |H|]$ and $j \in [1, |W|]$, that represents the suit-

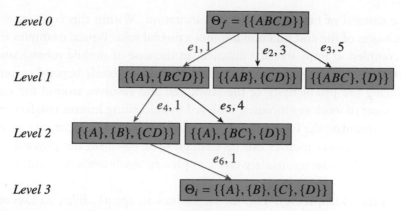

Figure 8.2 Example of auxiliary tree for a general assembly ABCD, highlighting the optimal path from Θ_f to Θ_i. The nodes are those visited by AO*.

ability of w_j in performing such a sequential assembly operation. If a specific action cannot be executed by one of the involved agents or results in being unsafe or time-consuming, to prevent the agent from being assigned to the action, we prune the corresponding hyperarc. By exploiting an optimality-based search algorithm, the path with the minimum total cost can be found. In this framework, costs are updated according to the human state monitored during the cooperation, ensuring an optimal role assignment at each assembly step.

Optimal AND/OR Graph Search. The AO* search algorithm draws inspiration from A* (Hart et al., 1968) and is tailored to function within the framework of an AND/OR graph. To extract the intended sequential assembly sequence and determine the optimal allocation of actions, a customized AO* search algorithm is introduced. It differs from the conventional AO* algorithm, which typically aims to minimize the path cost from the root node to either a single leaf node or a pair of leaf nodes (Martelli and Montanari, 1973). Instead, this algorithm is designed with the specific objective of examining all the leaf nodes, each of which corresponds to the initial configuration Θ_i of the hierarchical assembly. Consequently, the AO* search we employ continues until all leaf nodes, representing all assembly pieces within Θ_i, have been visited and incorporated into the path (Fig. 8.2). The implementation details of such an algorithm can be found in (Merlo et al., 2023).

Hyperarcs' Cost Computation. This framework retains a broad and versatile nature, offering the potential for various optimization approaches in the context of Human–Robot Collaboration. Within this context, the formulation of the cost function assumes a pivotal role. Typical examples are task completion time, traveled distance (in the case of mobile robots), and effort required. However, the objective of HRC extends beyond merely enhancing the productivity of the work cell and revolves around the enhancement of work ergonomics, achieved by mitigating human risk factors. To accommodate the inherent diversity of agents within collaborative work cells, three distinct metrics can be distinguished based on the physical attributes of these agents, namely task complexity, agent dexterity, and agent effort.

(i)   **Task Complexity.** This metric assesses an agent's ability to execute a given task in accordance with the specific requirements outlined by the production process. It accounts for the feasibility and implementation of various actions, acknowledging that certain actions may not be achievable by a collaborative robot (cobot) or might not be included in their repertoire.

(ii) **Agent Dexterity.** While task complexity evaluates the skills of individual agents in isolation, agent dexterity comes into play when multiple agents could potentially perform the same task. It is employed to prioritize agents with superior motion capabilities, effectively enhancing their suitability for specific tasks.

(iii) **Agent Effort.** This is another metric introduced to ensure equitable distribution of the physical efforts required among the agents. It serves to maintain a balance in the workload shared by different agents. Ergonomic evaluations, properly arranged for robotic agents (cf. (Merlo et al., 2023)), can also be included in this metric.

In addition, each of these metrics may be further broken down into one or more core components to provide a more comprehensive evaluation of the agents' capabilities and characteristics. A detailed explanation of those metrics and potential implementations are available in (Lamon et al., 2019).

8.1.3   Collaborative hierarchical assembly use case

The metric-driven allocation approach was subjected to validation within the context of a fast reconfigurable assembly line setup characterized by its flexibility. We recreated a collaborative hierarchical assembly task involving a mixed team of agents, whose roles were determined using the algorithm detailed earlier. In this proof-of-concept demonstration, we executed the physical assembly of a metallic structure, comprising two distinct aluminum profiles of varying weight and dimensions. These profiles were joined together via a corner joint, secured using screws and nuts. For the sake of clarity and brevity, the nomenclature for the objects in the subsequent graphs and tables has been abbreviated as follows: LP (Long Profile), SP (Small Profile), and CJ (Corner Joint).

The team consisted of a human worker and a collaborative Franka Emika Panda robot, equipped solely with its standard gripper. This configuration aligns with the principles of lean manufacturing, emphasizing the immediate utilization of existing resources over the development of highly customized tools. This approach is particularly suitable for scenarios involving small production batches, where the investment in specialized equipment may not be cost-effective.

To enhance the synchronization and coordination between these agents, each participant was equipped with a pair of Microsoft HoloLens, which incorporates mixed reality smart glasses technology. These devices are equipped with various sensors, including an Inertial Measurement Unit (IMU), a depth camera, a video camera, and microphones, facilitating natural interactions with the environment, holographic content, and contextual information. The HoloLens smartglasses feature a built-in gesture capture system and voice commands, which were leveraged to promote active collaboration among the agents. This technology enabled the display of the assembly status on the viewer and allowed the initiation of specific actions through the gesture capture system.

The experimental setup is illustrated in Fig. 8.3. The collaborative robot was positioned on a dedicated workbench, which served as the workspace for the assembly task. Meanwhile, the structural components required for the assembly were positioned on a separate storage desk. Essential tools for the task, including the Allen key, screws, and nuts, were prearranged and readily accessible on the workbench.

The determined sequence of actions essential for completing the assembly task, irrespective of the agents' skills, was as follows (please note that the order of actions is not unique):

1. Retrieve the small aluminum profile from the storage desk;
2. Position the small aluminum profile on the workbench;
3. Collect the corner joint from the storage desk;
4. Place the corner joint within the assembly workbench;
5. Insert a nut into the small profile and secure it to the corner joint, ensuring proper alignment of the screw with the nut. Tighten it using an Allen key;

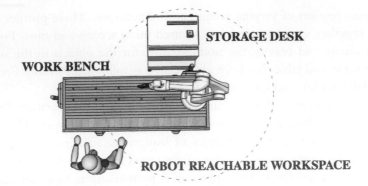

Figure 8.3 Setup of the experiment. The worker was outfitted with an augmented reality (AR) interface, with the purpose of ensuring and enhancing the worker's comprehension of the assigned tasks and the concurrent actions undertaken by the robot.

6. Acquire the long aluminum profile from the storage desk;
7. Set the long aluminum profile in the assembly area;
8. Insert a nut into the large profile and affix it to the opposite side of the corner joint, aligning the screw with the nut. Secure it with the Allen key.

In the initial three columns of Table 8.1, we have outlined the requisite skills and their feasibility for each agent involved. Notably, due to the absence of advanced perception capabilities and specialized tools for screwing, the actions labeled as "Align" and "Tool Action" were deemed infeasible for the cobot. However, all other actions remained feasible for both agents.

Table 8.1 Cost values for the aluminum profile assembly task; r denotes the robot, while h denotes the human.

Action	r	h	c_r	c_h	Assignment
Pickup + Transport + Place SP	✓	✓	1.360	∞	Cobot
Pickup + Transport + Place CJ	✓	✓	0.870	∞	Cobot
Align CJ with SP	✗	✓	∞	1.708	Human
Tool Action (Screwing)	✗	✓	∞	0.799	Human
Pickup + Transport + Place LP	✓	✓	0.842	∞	Cobot
Align LP with CJ	✗	✓	∞	1.736	Human
Hold CJ	✓	✓	0.313	0.454	Cobot
Tool Action (Screwing)	✗	✓	∞	1.427	Human

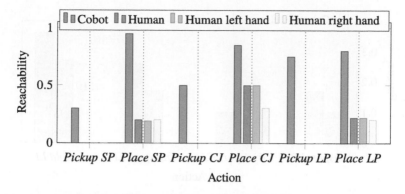

Figure 8.4 Reachability values computed for "Pickup" and "Place" skills.

The cost function values and role assignments depicted in Table 8.1, computed using the AO* algorithm, reveal the strategic allocation of tasks. The cobot's precision and power were harnessed in repetitive "Pick & Place" actions and in tasks requiring endurance to bear the weight of the objects. Conversely, the human agent's hand–eye coordination and task comprehension were pivotal for actions involving the alignment of screws with nuts.

To assess task assignments, we computed the cost function employing the proposed metrics for each action and each agent. For the sake of simplicity, we set all cost function weights $(\beta_{T_i}, \beta_{D_i}, \beta_{V_i})$ to 1. Task complexity metrics only necessitated knowledge of the agent's skill set (as specified in Table 8.1. In contrast, the agent dexterity metric required additional inputs, such as the agent's kinematic information and inverse kinematics constraints, along with action-related parameters like spatial positioning.

Specifically, Fig. 8.4 illustrates the reachability index (D) values for "Pickup" and "Place" skills concerning the small profile, corner joint, and long profile. Notably, the human agent's "Pickup" action had a reachability index of 0 due to the placement of the storage desk outside the worker's workspace. Consequently, based on this metric, the cobot was the most suitable agent for performing both the "Pickup" and "Place" actions.

On the other hand, the agent effort metric considered factors such as load weight, nominal execution time for each action, and the agent's joint capacity vector (**C**). In our experiment, we maintained consistent values for the joints of each agent, with $C_{cobot} = 100$ and $C_{human} = 50$, reflecting the cobot's capacity to exert higher torques with less fatigue compared to the

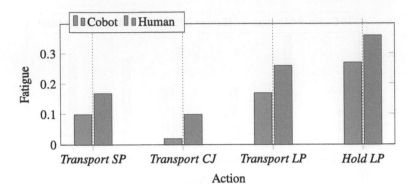

Figure 8.5 Fatigue values computed with for "Transport" and "Hold" skills.

human agent. This is depicted in Fig. 8.5, which shows the fatigue levels. Notably, we focused on fatigue stemming from external load handling due to the absence of a comprehensive dynamic model for the human agent. Fatigue metrics were computed exclusively for actions that involved load carrying, such as "Transport" and "Hold." The nominal execution time for tasks was set at 20 seconds for the "Transport" action involving the three objects and 60 seconds for the "Hold" action with the long profile.

The computed fatigue indices did not account for previously accumulated fatigue levels. However, the broader approach should consider the fatigue linked to prior actions, as outlined in the algorithm by Lamon et al. (2018).

8.2 Interface (MOCA-MAN): Human–robot comanipulation

8.2.1 Connotation

Interfaces play a crucial role in human–robot interaction and collaboration due to their ability to bridge the gap between the inherently distinct modalities and cognitive processes of humans and robots. These interfaces facilitate the exchange of information and commands between the two entities, allowing humans to communicate their intentions and preferences while enabling robots to perceive and respond effectively (Ajoudani et al., 2018). The design and functionality of interfaces directly impact the usability, efficiency, and safety of such interactions. Moreover, interfaces serve as a means to abstract complex robot capabilities into user-friendly control mechanisms, making advanced robotic systems accessible to a broader user base. Through interfaces, researchers and engineers can also leverage insights from fields such as human–computer interaction and cognitive psychology to improve the overall quality of human–robot interactions and collaboration, promoting collaboration and cooperation in various application domains, from healthcare and manufacturing to autonomous vehicles and entertainment.

In terms of robotic manipulators, for fixed-base manipulators, it is easy to achieve human–robot physical interaction by gravity compensation. In this way, humans can operate the robots freely by direct kinesthetic teaching. However, this method is unsuitable for a Collaborative Mobile Manipulator (CMM). A collaborative mobile manipulator is a versatile robotic system that combines the mobility of a mobile platform with the dexterity of a robotic manipulator, typically an articulated collaborative robotic arm with an end effector. This integration allows the robot to move and manipulate objects in complex, dynamic environments. Mobile manipulators are of paramount importance in robotics research due to their multifaceted capabilities and broad range of applications. They are essential for tasks such as warehouse automation, search and rescue missions, and home assistance. Their significance lies in their potential to address real-world challenges by autonomously navigating to a target location and performing intricate manipulation tasks, all while adapting to changing environments and interacting with humans (Zhao et al., 2022). The research on mobile manipulators is crucial in advancing robotics technology, as it explores topics like perception, control, planning, and human–robot interaction and collaboration, pushing the boundaries of what robots can accomplish in both structured and unstructured settings.

To address the above issues, we designed an interface for mobile manipulators, namely MOCA-MAN (Kim et al., 2020; Gandarias et al., 2022), which is designed for our MObile Collaborative robotic Assistant (MOCA, see Fig. 8.6) (Wu et al., 2019), allowing humans to operate the mobile manipulators intuitively. The details of this interface are presented in Section 8.2.3.

8.2.2 MOCA platform and whole-body impedance control

MOCA (as illustrated in Fig. 8.6) constitutes a research-oriented robotic platform meticulously designed to facilitate Human–Robot Collaboration. Its loco-manipulation capabilities make it a promising candidate for various applications, including logistics (Lamon et al., 2020b; Balatti et al.,

End-Effector

Franka Emika Panda

Robotnik SUMMIT-XL STEEL

Figure 8.6 MOCA platform: the Franka Emika Panda robotic arm is integrated on the top of the Robotnik SUMMIT-XL STEEL mobile platform. The End-effector can be changed according to different tasks.

2020a) and flexible manufacturing (Fusaro et al., 2021b). The system is composed of a lightweight torque-controlled 7-DoFs Franka Emika Panda robotic arm, mounted on top of the velocity-controlled 3-DoFs Robotnik SUMMIT-XL STEEL mobile platform.

In general, CMM control strategies can be divided into two main types, independent and whole-body. The independent type consists of two separate controllers, one for the arm and one for the mobile platform. Such a control architecture can be implemented faster, but it suffers from synchronization issues since each controller ignores the dynamics coupling between the two independent components (Ellekilde and Christensen, 2009). For these reasons, such a strategy forces the arm and the base to move alternately. Hence, the loco-manipulation capabilities of the platform cannot be fully exploited.

In order to address the synchronization challenges inherent in disentangled strategies, comprehensive whole-body controllers have been introduced. Specifically, torque-based whole-body strategies have emerged as a solution capable of not only governing interaction forces with the environment by adopting impedance-like characteristics but also managing the distribution of motion at the joint level. This results in the generation of diverse motion patterns encompassing manipulation, locomotion, and loco-manipulation (Lamon et al., 2020b; Wu et al., 2021).

The details of the proposed weighted whole–body impedance controller in (Lamon et al., 2020b) for MOCA are presented as follows. The whole-body dynamic model can be formulated as:

$$
\begin{pmatrix} M_v & 0 \\ 0 & M_a(q_a) \end{pmatrix} \begin{pmatrix} \ddot{q}_m \\ \ddot{q}_a \end{pmatrix} + \begin{pmatrix} D_v & 0 \\ 0 & C_a(q_a, \dot{q}_a) \end{pmatrix} \begin{pmatrix} \dot{q}_m \\ \dot{q}_a \end{pmatrix} \\ + \begin{pmatrix} 0 \\ g_a(q_a) \end{pmatrix} = \begin{pmatrix} \tau_m^{vir} \\ \tau_a \end{pmatrix} + \begin{pmatrix} \tau_m^{ext} \\ \tau_a^{ext} \end{pmatrix},
\tag{8.1}
$$

where $M_v \in \mathbb{R}^{n_b \times n_b}$ and $D_v \in \mathbb{R}^{n_b \times n_b}$ are the virtual inertia and virtual damping of the mobile platform, $\dot{q}_m \in \mathbb{R}^{n_b}$ is its input velocity, $\tau_m^{ext} \in \mathbb{R}^{n_b}$ and $\tau_m^{vir} \in \mathbb{R}^{n_b}$ are the related external and the virtual torque. With respect to the arm, $q_a, \dot{q}_a$, and $\ddot{q}_a \in \mathbb{R}^{n_a}$ are the joint angles, velocities and accelerations vectors, $M_a \in \mathbb{R}^{n_a \times n_a}$ is the symmetric and positive definite inertia matrix of the arm, $C_a \in \mathbb{R}^{n_a}$ is the Coriolis and centrifugal force vector, $g_a \in \mathbb{R}^{n_a}$ is the gravity, $\tau_a \in \mathbb{R}^{n_a}$, and $\tau_a^{ext} \in \mathbb{R}^{n_a}$ are the commanded torque vector and external torque vector, respectively. This model can be summarized by

$$
M(q)\ddot{q} + C(q, \dot{q})\dot{q} + g(q) = \tau^u + \tau^{ext},
\tag{8.2}
$$

where $M(q) \in \mathbb{R}^{n \times n}$ $(n = n_a + n_b)$ is the symmetric positive definite joint-space inertia matrix, $C(q, \dot{q}) \in \mathbb{R}^{n \times n}$ is the joint-space Coriolis/centrifugal matrix, and $g(q) \in \mathbb{R}^n$ the joint-space gravity. Finally, $\tau^u \in \mathbb{R}^n$ and $\tau^{ext} \in \mathbb{R}^n$ represent joint-space input and external torque.

The MOCA Cartesian impedance controller is formulated as a prioritized weighted inverse dynamics algorithm and can be obtained by solving the problem of finding the torque vector τ closest to some desired τ_0 that realizes the operational forces $F \in \mathbb{R}^m$, according to the norm induced by the positive definite weighting matrix $W \in \mathbb{R}^{n \times n}$,

$$
\min_{\tau \in \mathbb{R}^n} \frac{1}{2} \|\tau - \tau_0\|_W^2 \quad \text{such that } F = \bar{J}^T \tau,
\tag{8.3}
$$

where $\bar{J} = M^{-1}J^T \Lambda$ is the dynamically consistent pseudoinverse of the Jacobian matrix $J(q)$, and the constraint $F = \bar{J}^T \tau = \Lambda J M^{-1} \tau$ is the general relationship between the generalized joint torques and the operational forces, and $\Lambda = (JM^{-1}J^T)^{-1} \in \mathbb{R}^{m \times m}$ is the Cartesian inertia. The closed-form solution results in:

$$
\tau^u = W^{-1}M^{-1}J^T \Lambda_W \Lambda^{-1} F + (I - W^{-1}M^{-1}J^T \Lambda_W J M^{-1})\tau_0,
\tag{8.4}
$$

where $\mathbf{\Lambda}_W = (\mathbf{J}\mathbf{M}^{-1}\mathbf{W}^{-1}\mathbf{M}^{-1}\mathbf{J}^T)^{-1}$ is the weighted Cartesian inertia, analogous to the Cartesian inertia $\mathbf{\Lambda}$. The weighting matrix $\mathbf{W}$ is generally defined as $\mathbf{W}(\mathbf{q}) = \mathbf{H}^T\mathbf{M}^{-1}(\mathbf{q})\mathbf{H}$, where $\mathbf{H} \in \mathbb{R}^{n \times n}$ is the tunable positive definite weight matrix of the controller, that is used to generate different motion modes (Lamon et al., 2020b).

Finally, $\mathbf{F}$ is computed to generate the desired closed-loop behavior, according to the Cartesian impedance law,

$$\mathbf{F} = -\mathbf{D}^d\dot{\mathbf{x}} - \mathbf{K}^d\tilde{\mathbf{x}}, \tag{8.5}$$

where $\tilde{\mathbf{x}} = \mathbf{x}_d - \mathbf{x} \in \mathbb{R}^6$ is the Cartesian position error computed with respect to the desired Cartesian pose $\mathbf{x}_d$ (analogously for its derivatives $\dot{\tilde{\mathbf{x}}}$ and $\ddot{\tilde{\mathbf{x}}}$), and $\mathbf{K}^d \in \mathbb{R}^{m \times m}$ and $\mathbf{D}^d \in \mathbb{R}^{m \times m}$ are the desired Cartesian inertia, damping, and stiffness matrices, respectively. Moreover, $\boldsymbol{\tau}_0$ could contain different contributions, such as joint impedance, collision and self-collision avoidance, joint limits avoidance, etc.

8.2.3 Design and control of the interface

The overall design of the MOCA-MAN interface can be found in Fig. 8.7. It consists of the following items:

1. An Arduino Nano microcontroller connected to a 4-button panel allows the user to configure different functionalities and communicate with the robot through the ROS middleware suite.
2. A force–torque (F/T) sensor to measure the user interaction wrenches with a physical part that the human can easily grasp.
3. An End-Effector (EE) connector connecting the interface and the robotic arm EE.

The interface is connected to the EE of the robotic manipulator. Then, the operators can hold the handler to control MOCA. The measured human wrenches $\hat{\boldsymbol{\lambda}}_h \in \mathbb{R}^m$, $m \leq 6$ are used to change the input equilibrium pose $(\mathbf{x}_d)$ in (8.5) as input of an admittance controller which implements the following desired dynamics:

$$\mathbf{M}^{adm}\ddot{\mathbf{x}}_d + \mathbf{D}^{adm}\dot{\mathbf{x}}_d = \hat{\boldsymbol{\lambda}}_h, \tag{8.6}$$

where $\mathbf{M}^{adm}, \mathbf{D}^{adm} \in \mathbb{R}^{m \times m}$ are respectively the admittance mass and damping. This way, MOCA can move in space in the same direction as the applied wrench, and the operators can control MOCA freely. In this way,

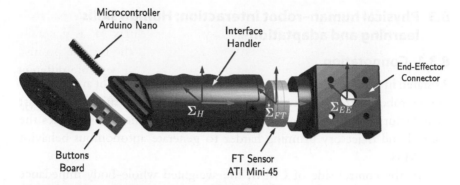

Microcontroller
Arduino Nano

Interface
Handler

End-Effector
Connector

Σ_H Σ_{FT} Σ_{EE}

Buttons
Board

FT Sensor
ATI Mini-45

Figure 8.7 Overall design of MOCA-MAN interface. The interface is connected to the robotic arm EE by the connector. The humans can hold the handler to move the mobile manipulator.

operators can focus on high-level actions, for example, target location and obstacle avoidance, and MOCA provides physical assistance, i.e., holding heavy objects. This human–robot comanipulation manner combines the advantages of both sides, namely, the human's decision and perception intelligence and the robot's physical assistance capability, improving the efficiency of the tasks and ergonomics of the human (Lamon et al., 2020a).

The integration of the MOCA-MAN interface introduces supplementary functionalities that enable the optimal utilization of a mobile-based robot, streamlining the process of task comanipulation while accommodating the variable preferences inherent in human-guided behaviors. By engaging the loco-manipulation mode, the human instructor gains the capability to navigate the robot across a theoretically boundless workspace during the human–robot comanipulation while maintaining a stationary, mobile base when planning interactions between the end-effector and the environment. Notably, given that mobile base movements, operating at a lower frequency compared to the arm can introduce perturbations to the end-effector's trajectory, presenting a predefined interaction with the environment can present challenges. Subsequently, the human instructor retains the ability to fine-tune the system's responsiveness to the interaction by selecting the admittance based on their experience, preferences, and specific task requirements, opting for high admittance for extensive, unconstrained motions and low admittance for concise, precision-based interactions with the environment. Furthermore, the selective activation of a subset of task space axes proves instrumental in task execution.

8.3 Physical human–robot interaction: Human skills' learning and adaptation

8.3.1 Connotation

As stated in Section 8.2, CMM is integrated by a mobile robot and a collaborative robotic arm. Although it makes the CMM more suitable for smart manufacturing scenarios, integrating different components also makes the control and trajectory planning harder to generate autonomous behavior for CMM.

For the control side of CMM, the weighted whole-body impedance controller presented in Section 8.2.2 is a promising solution when the robot interacts with the environment. While characterizing the system's response to interactions with the environment, impedance controllers necessitate the adjustment of a substantial number of design parameters, typically encompassing stiffness, damping, and inertia. The choice of appropriate impedance levels hinges on the specific task requirements; higher impedance values are desirable for precise motions, while lower impedance settings are advantageous for facilitating interactions with the environment. However, the optimal tuning of these impedance parameters, often referred to as Variable Impedance Control (VIC), remains an ongoing challenge in the field.

For example, Duan et al. (2018) utilized an impedance model based on a mass–damper system to articulate the desired dynamics of the positional error between the anticipated environment position and the actual desired position, which corresponds to force tracking. In this approach, the damping is adaptively adjusted based on the force error, and the resultant dynamic model is used to command a position-controlled robot. While this method demonstrated the capability to achieve force tracking on diverse surfaces, it exhibited limitations when the surface contact was unexpectedly lost, potentially leading to precarious robot behavior.

In contrast, alternative strategies put forth by Averta and Hogan (2020); Balatti et al. (2020b) propose self-tuning mechanisms for impedance parameters contingent upon positional error. These strategies entail an increment in stiffness in response to heightened tracking error, while the interaction force is leveraged to detect unanticipated contacts and restore a compliant behavior. Furthermore, Averta and Hogan (2020) attempted to minimize interaction with the environment; however, it was acknowledged that certain tasks necessitate high interaction forces, rendering compliance unsuitable.

Human-in-the-loop approaches offer the possibility of online adjustments to the robot's behavior, allowing for dynamic variations in damping and inertia based on human input, thereby achieving a balance between precision and task execution speed for collaborative endeavors (Ficuciello et al., 2015). Some methods even incorporate EMG sensors to estimate joint stiffness and simultaneously generate reference trajectories (Peternel et al., 2014; Wu et al., 2020). Nevertheless, it is important to note that these human impedance-transfer techniques, particularly those applied to fixed-base robotic arms, require real-time adaptation and are most applicable in contexts involving teleoperation. However, their reliance on EMG sensors and motion capture systems renders them less suitable for widespread adoption in industrial and domestic settings.

Moreover, in the context of generating autonomous tasks for CMM, it is imperative not only to establish optimal impedance profiles but also to define reference trajectories. The IL paradigm has emerged as a highly effective approach for acquiring desired motion patterns, utilizing a predetermined set of trajectories demonstrated by an instructor, as evidenced by prior research (Calinon and Lee, 2019). This paradigm encompasses a range of algorithms for encoding these demonstrated motions, with prominent methods including Dynamical Movement Primitives (DMPs) (Ijspeert et al., 2013) and GMM/Gaussian Mixture Regression (GMR) (Calinon and Lee, 2019), among others. In the context of Collaborative Mobile Manipulators, one specific IL methodology known as "kinesthetic teaching," namely, the human teaching the robot directly by physical interaction, has gained substantial traction due to its distinct advantage of obviating the need for retargeting from the human instructor to the robot. Implementation of this approach typically necessitates mechanisms for gravity compensation or admittance control, often achieved through the utilization of force sensing technologies, such as Force–Torque (F/T) sensors located at the end-effector or joint torque sensors.

To the best of the authors' knowledge, the current landscape reveals a scarcity of intuitive interfaces that effectively harness the potential of CMMs in the context of kinesthetic teaching. Within the domain of fixed-base manipulators, several studies have grappled with the intricate challenge of concurrent learning of trajectories and force profiles through impedance models (Abu-Dakka and Saveriano, 2020), employing GMM/GMR. Notably, in the work by Calinon et al. (2010), a strategy was devised to generate robot behavior while ensuring safety, involving the fine-tuning of variable stiffness based on the inverse of the position covariance of the

GMM. Furthermore, in the study conducted by Abu-Dakka et al. (2018), researchers explored two distinct Semipositive Definite (SPD) stiffness representations. Stiffness profiles were obtained via a least-squares estimator using a linear interaction model, drawing on position and force demonstrations.

A similar approach, underpinned by regularized regression and force sensing, was applied by Michel et al. (2021), where the GMM established a mapping between external force (input) and stiffness (output). Consequently, stiffness was generated in real-time through GMR based on the prevailing external force during teleoperation, leading to a direct correlation between high external force and elevated stiffness levels. However, it is worth noting that the substantial stiffness of robots may pose risks to humans when subject to substantial disturbances caused by human interactions.

Alternatively, a distinct methodology rooted in RL was developed, where the robot policy was encapsulated by a DMP encompassing trajectory and impedance parameters. Subsequently, policy parameters were subject to offline optimization employing a variant of policy improvement with path integrals (Buchli et al., 2011). Nevertheless, it is important to acknowledge that this RL approach and the iterative optimization method introduced by Averta and Hogan (2020) necessitate multiple simulated task executions and may not adequately address unforeseen disturbances in real-time settings. Furthermore, it is noteworthy that a considerable portion of the existing literature lacks explicit provisions for ensuring the stability of VIC, a fundamental attribute for fostering safe physical Human–Robot Interaction (pHRI) (Abu-Dakka and Saveriano, 2020).

In response to these challenges, we present a novel approach aimed at instructing locomotion and manipulation tasks, capitalizing on the VIC paradigm. Our primary objective is to facilitate the training of our MOCA through human physical demonstrations, employing IL techniques. The ultimate goal is to generate robust and secure autonomous behaviors that meticulously regulate the interaction forces occurring between the robot and the complex, unstructured environment, particularly under quasistatic conditions. The methodology leverages the MOCA-MAN interface, as presented in Section 8.2, which empowers the intuitive demonstration of intricate interactive tasks. These tasks are subsequently encoded using a GMM, enabling the simultaneous acquisition of task trajectory and interaction force maps between the robot and its environment. The desired trajectory and force profiles are synthesized utilizing GMR. The computation of desired impedance parameters, contingent upon the learned force, occurs online via a Quadratic Program (QP), ensuring the minimal stiffness necessary to execute the task while adhering to predefined force and stiffness constraints mandated by safety considerations. Furthermore, an energy tank-based passivity constraint is incorporated to bolster the stability of the controller. The synthesized desired trajectory and stiffness values are then dispatched to the MOCA's weighted whole-body impedance controller, as depicted in Fig. 8.8.

We validate the effectiveness of our method through a table cleaning task, conducting two distinct experiments. Initially, the task is instructed under nominal conditions, after which MOCA endeavors to replicate the acquired knowledge in an identical setting while also contending with external disturbances originating from the environment and unforeseen physical interactions with humans. The results demonstrate that our approach to impedance parameter tuning consistently yields commendable performances across all conditions. It facilitates simultaneous trajectory and force tracking, fostering robust and compliant interactions, and notably outperforming constant stiffness solutions.

8.3.2 Methodology

8.3.2.1 Desired trajectory and interaction force hybrid learning exploiting MOCA-MAN interface

The MOCA-MAN interface is applied for human teachers to demonstrate the desired end-effector trajectories and interaction wrenches with MOCA. We integrated kinesthetic teaching with the MOCA-MAN interface. The overall setup is depicted in Fig. 8.9.

Throughout the demonstration phase, the aforementioned interface elements serve as instrumental tools, affording the human instructor control over various functionalities within the admittance mapping, as outlined in (8.6), and the overarching whole-body controller detailed in Section 8.2.2. Specifically, the human operator exercises influence over the locomotion and manipulation behavior of the robot, a capability enacted through the manipulation of parameters τ_0 and H, as well as over the specified admittance characteristics represented by M^{adm} and D^{adm}. Furthermore, the interface provides the option to activate or deactivate rotations and translations of the end-effector.

Moreover, the system's responsiveness to interactions can be fine-tuned by the human instructor, allowing for the selection of an appropriate ad-

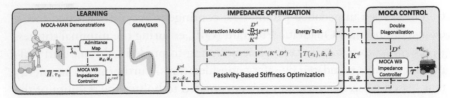

Figure 8.8 The framework, as delineated in the proposed scheme, enables direct human instruction of MOCA in interactive tasks through the MOCA-MAN interface. Subsequently, the desired trajectories and interaction forces are faithfully reproduced through GMM and GMR. Following this, the desired force information is channeled into a QP-based algorithm for the real-time optimization of stiffness parameters. To uphold the system's passivity, an energy tank constraint is imposed. Ultimately, the desired trajectory and stiffness values are conveyed to MOCA's comprehensive whole-body impedance controller, affecting the execution of the instructed tasks with precision and compliance.

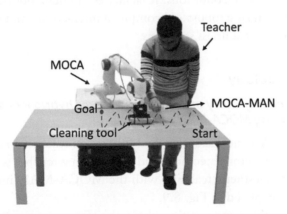

Figure 8.9 Experimental setup. The teacher demonstrates the table cleaning task to MOCA-MAN from Start to Goal. A cleaning tool is attached to the end-effector of MOCA.

mittance setting tailored to their experience, preferences, and the precise requirements of the task at hand. High admittance values prove advantageous for facilitating extended, unconstrained motions, while low admittance settings are better suited for brief and precise interactions with the environment. Additionally, the capacity to select specific subsets of task space axes may streamline task execution and expedite the learning process, further enhancing the versatility of the interface.

The desired pose $\boldsymbol{x}_d$ and twist $\dot{\boldsymbol{x}}_d$ are derived as solutions to the equation presented in (8.6). As the robotic arm under consideration operates under torque control, it affords the capability to estimate the interaction wrenches occurring at the robot's end-effector interface with the surrounding environment, denoted as $\hat{\boldsymbol{F}}^{ext}$. Consequently, during the demonstration phase, the human instructor is enabled to provide guidance to the robot via the admittance-type physical interface, while concurrently the robot arm autonomously estimates and responds to its interactions with the environment. This real-time interaction monitoring and response mechanism enhances the system's adaptability and responsiveness during the learning process.

The desired trajectories and force profiles are generated using GMM and GMR. A GMM is defined as a linear superposition of several Gaussian distributions,

$$p(\boldsymbol{\xi}) = \sum_{k=1}^{K} \pi_k \mathcal{N}(\boldsymbol{\xi}|\boldsymbol{\mu}_k, \boldsymbol{\Sigma}_k), \tag{8.7}$$

where $\boldsymbol{\xi} \in \mathbb{R}^d$ and $p(\boldsymbol{\xi}) \in \mathbb{R}$ represent the vector of variables (input and output variables) and joint probability distribution, respectively; $\pi_k \in \mathbb{R}$, $\boldsymbol{\mu}_k \in \mathbb{R}^d$, and $\boldsymbol{\Sigma}_k \in \mathbb{R}^{d \times d}$ are the prior probability, mean, and covariance of the k-th Gaussian component, respectively. And $K \in \mathbb{Z}^+$ is the number of Gaussian distributions.

The parameters of GMM can be estimated by Expectation Maximization (EM) algorithm (Bishop and Nasrabadi, 2006) with an offline training process that makes use of the demonstrations. To make the representation easier to understand, we give the following definition: $\boldsymbol{\eta}^{\mathcal{I}}$ and $\boldsymbol{\eta}^{\mathcal{O}}$ respectively denote the input and output variables on which the training is carried out, where the subscripts $\mathcal{I}$ and $\mathcal{O}$ mean their corresponding dimensions. Based on these, the generic data point $\boldsymbol{\eta} \in \mathbb{R}^d$, the mean $\boldsymbol{\mu}_k$, and covariance $\boldsymbol{\Sigma}_k$ of the k-th Gaussian component can be written as follows:

$$\boldsymbol{\eta} = \begin{bmatrix} \boldsymbol{\eta}^{\mathcal{I}} \\ \boldsymbol{\eta}^{\mathcal{O}} \end{bmatrix}, \quad \boldsymbol{\mu}_k = \begin{bmatrix} \boldsymbol{\mu}_k^{\mathcal{I}} \\ \boldsymbol{\mu}_k^{\mathcal{O}} \end{bmatrix}, \quad \boldsymbol{\Sigma}_k = \begin{bmatrix} \boldsymbol{\Sigma}_k^{\mathcal{II}} & \boldsymbol{\Sigma}_k^{\mathcal{IO}} \\ \boldsymbol{\Sigma}_k^{\mathcal{OI}} & \boldsymbol{\Sigma}_k^{\mathcal{OO}} \end{bmatrix}. \tag{8.8}$$

Given a input variable $\boldsymbol{\eta}^{\mathcal{I}}$, the best estimation of output $\hat{\boldsymbol{\eta}}^{\mathcal{O}}$ is the mean $\hat{\boldsymbol{\mu}}$ of the conditional probability distribution $\hat{\boldsymbol{\eta}}^{\mathcal{O}}|\boldsymbol{\eta}^{\mathcal{I}} \sim \mathcal{N}(\hat{\boldsymbol{\mu}}, \hat{\boldsymbol{\Sigma}})$, which can be obtained by (Huang et al., 2019):

$$\hat{\mu} = \mathbb{E}(\hat{\eta}^{\mathcal{O}}|\eta^{\mathcal{I}}) = \sum_{k=1}^{K} h_k(\eta^{\mathcal{I}})\mu_k(\eta^{\mathcal{I}}), \tag{8.9}$$

where

$$h_k(\eta^{\mathcal{I}}) = \frac{\pi_k \mathcal{N}(\eta^{\mathcal{I}}|\mu_k^{\mathcal{I}}, \Sigma_k^{\mathcal{II}})}{\sum_{j=1}^{K} \pi_j \mathcal{N}(\eta^{\mathcal{I}}|\mu_j^{\mathcal{I}}, \Sigma_j^{\mathcal{II}})}, \tag{8.10}$$

$$\mu_k(\eta^{\mathcal{I}}) = \mu_k^{\mathcal{O}} + \Sigma_k^{\mathcal{OI}}(\Sigma_k^{\mathcal{II}})^{-1}(\eta^{\mathcal{I}} - \mu_k^{\mathcal{I}}). \tag{8.11}$$

Here, we set the input variable as time ($\eta^{\mathcal{I}} = t$), and the output variables are pose, twist, and wrench $\left(\eta^{\mathcal{O}} = \begin{bmatrix} x_d^T & \dot{x}_d^T & \hat{F}^{ext^T} \end{bmatrix}^T \right)$.

8.3.2.2 QP-based stiffness online optimization

The outcomes of the GMR model serve as inputs to a QP algorithm, which facilitates the real-time adjustment of stiffness parameters within the Cartesian whole-body impedance controller, as elaborated upon in Section 8.2.2. The QP is formulated as follows:

$$\min_{\substack{K_i^d \in \mathbb{R}^{m \times m} \\ i \in \{1, \dots, T\}}} \frac{1}{2} \sum_{i=1}^{T} \left(\|F_i^{ext} - F_i^d\|_Q^2 + \|K_i^d - K^{min}\|_R^2 \right)$$

$$\text{such that} \quad K^{min} \leq K_i^d \leq K^{max}, \quad i \in \{1, \dots, T\}, \tag{8.12}$$

$$-F^{max} \leq F_i^{ext} \leq F^{max}, \quad i \in \{1, \dots, T\},$$

where i is the time step, T is the length of the time window, Q and R $\in \mathbb{R}^{m \times m}$ are diagonal positive definite weighting matrices, $K_i^d \in \mathbb{R}^{m \times m}$ is the desired stiffness of the Cartesian whole body impedance controller at time step i, K^{min} and $K^{max} \in \mathbb{R}^{m \times m}$ are respectively the minimum and maximum allowed stiffnesses, $F_i^{ext} \in \mathbb{R}^m$ is the wrench of the impedance interaction model at time step i, that can be modeled with impedance-like laws, $F_i^d \in \mathbb{R}^m$ is the learned desired interaction wrench at time step i, and $F^{max} \in \mathbb{R}^m$ is the maximum wrench that the robot can exert. The constraint inequality applied to vector elements is performed on an element-wise basis. The optimization problem delineated in the preceding sections encompasses a trade-off between achieving precise tracking of the desired wrench while concurrently imposing constraints to maintain low stiffness levels. This trade-off is essential for optimizing both control performance and safety in the system.

The desired impedance interaction model can be expressed in different possible ways. To render a desired mass–spring–damper system, F^{ext} is expressed as

$$F^{ext} = \Lambda^d \ddot{\tilde{x}} + D^d \dot{\tilde{x}} + K^d \tilde{x}, \qquad (8.13)$$

where $\Lambda^d \in \mathbb{R}^{m \times m}$ is the desired inertia matrix. However, the interaction wrench must be measured precisely to render the desired inertia matrix, which is referred to as inertia shaping (e.g., using an F/T sensor). Unfortunately, a precise measure is often not available, and the following interaction model is rendered (Ott, 2008):

$$F^{ext} = \Lambda(x) \ddot{\tilde{x}} + \big(\mu(x, \dot{x}) + D^d\big) \dot{\tilde{x}} + K^d \tilde{x}, \qquad (8.14)$$

where no inertia shaping is performed, since the actual Cartesian inertia of the manipulator $\Lambda(x)$ is used. Note that, in order to keep the physical coherence of the interaction model, the Coriolis and centrifugal terms $\mu(x, \dot{x})$ must be added to the desired damping since those arise from a configuration dependent inertia.

In practice, since also the acceleration signal is often noisy, the following simplified model is used:

$$F^{ext} = D^d \dot{\tilde{x}} + K^d \tilde{x}. \qquad (8.15)$$

8.3.2.3 Tank energy based passivity constraint

The QP framework, as expounded upon in Section 8.3.2.2, computes the Cartesian stiffness on a per-time-step basis, which is subsequently transmitted to the Cartesian impedance whole-body controller, elucidated in Section 8.2.2. It is a well-established understanding that variable impedance controllers may introduce the risk of breaching system passivity, potentially compromising the system's stability (Ferraguti et al., 2015; Hjorth et al., 2023). Consequently, the stability of the controlled system cannot be unequivocally guaranteed. To rectify this, the previously introduced QP formulation can be augmented with the incorporation of a passivity constraint specifically designed for the power port $\dot{x} F^{ext}$.

The interaction model characterizing the variable Cartesian impedance can be represented as a port-Hamiltonian system, and this model can be further enhanced by the inclusion of an energy tank (Ferraguti et al., 2015).

This augmentation fortifies the system's passivity and ensures that it remains stable, even when subjected to variable impedance control strategies, providing a robust and reliable control framework. The scalar differential equation that describes the tank dynamics is

$$\dot{x}_t = \frac{\sigma}{x_t}\dot{\tilde{x}}^T D^d \dot{\tilde{x}} - \frac{w^T}{x_t}\dot{\tilde{x}}, \tag{8.16}$$

where $x_t \in \mathbb{R}$ is the state of the tank that stores energy $T(x_t) = \frac{1}{2}x_t^2$, $\sigma \in \{0, 1\}$ is used to enable and disable the dissipated energy storage in case a maximum limit is reached, and w is the extra input of the port-Hamiltonian dynamics which can be written as

$$w(t) = \begin{cases} -K^v(t)\tilde{x} & \text{if } T(x_t) > \varepsilon, \\ 0 & \text{otherwise,} \end{cases} \tag{8.17}$$

where $K^v(t)$ is the variable part of the stiffness so that $K^d(t) = K^{min} + K^v(t)$ and $\varepsilon > 0$ is the minimum energy that the tank is allowed to store. The tank energy is initialized so that $T(x_t(0)) > \varepsilon$. The constraint used to augment the QP is derived by integrating the energy tank over time and enforcing it to be higher than its minimum ε,

$$T(x_t(t)) = T(x_t(t-1)) + \dot{T}(x_t(t))\Delta t > \varepsilon, \tag{8.18}$$

where Δt is the time step and

$$\dot{T}(x_t) = x_t\dot{x}_t = \sigma\dot{\tilde{x}}^T D^d \dot{\tilde{x}} - w^T\dot{\tilde{x}} \tag{8.19}$$

$$= \begin{cases} \sigma\dot{\tilde{x}}^T D^d \dot{\tilde{x}} + \tilde{x}^T K^v \dot{\tilde{x}} & \text{if } T(x_t) > \varepsilon, \\ \sigma\dot{\tilde{x}}^T D^d \dot{\tilde{x}} & \text{otherwise.} \end{cases} \tag{8.20}$$

Note from (8.20) that the constraint can be expressed as a linear function of the QP optimization variable only when $T > \varepsilon$. Instead, when $T \leq \varepsilon$ the stiffness is constrained to take on its minimum value ($K^d = K^{min}$).

8.3.2.4 Technical details

In the preceding subsection, we provided a comprehensive explanation of the method, yet it is imperative to acknowledge that certain simplifying assumptions were made in its implementation. First and foremost, the QP optimization incorporates an interaction model as delineated in (8.15), wherein the matrices K^d and D^d are presumed to be diagonal. Furthermore, the optimization procedure exclusively addresses the translational components of the interaction model, while the rotational aspects remain held constant. The value of $\boldsymbol{D}^d$ in the optimization at the subsequent control loop is derived through double diagonalization, as previously established in the optimization step for $\boldsymbol{K}^d$, with consideration for a critical damping factor (Ott, 2008). The diagonal elements involved in this computation are as follows:

$$\boldsymbol{d}^d(t) = 2 \cdot 0.707 \cdot \sqrt{\boldsymbol{k}^d(t-1)}, \tag{8.21}$$

where $\boldsymbol{d}^d(t) \in \mathbb{R}^3$ is the vector of the diagonal components of the desired Cartesian damping matrix at time t and $\boldsymbol{k}^d(t-1) \in \mathbb{R}^3$ is the vector of the diagonal components of the desired Cartesian stiffness matrix at time $t-1$. Damping is calculated based on the desired stiffness from the preceding time step to uphold the quadratic characteristics of the cost function within the QP formulation. It is worth noting that the current approach involves a one-time step consideration in each optimization ($T = 1$); however, it is conceivable that employing larger values of T may yield more gradual and smoother optimization solutions. This presents an avenue for future exploration and potential refinement of the optimization process.

The problem in (8.12) can be rewritten as:

$$\min_{\boldsymbol{k}^d \in \mathbb{R}^3} \frac{1}{2} \left(\|\boldsymbol{F}^{ext} - \boldsymbol{F}^d\|_{\boldsymbol{Q}}^2 + \|\mathrm{diag}\{\boldsymbol{k}^d\} - \boldsymbol{K}^{min}\|_{\boldsymbol{R}}^2 \right)$$
$$\text{such that} \quad \boldsymbol{k}^{min} \leq \boldsymbol{k}^d \leq \boldsymbol{k}^{max}, \tag{8.22}$$
$$-\boldsymbol{F}^{max} \leq \boldsymbol{F}^{ext} \leq \boldsymbol{F}^{max},$$
$$T(x_t) \geq \varepsilon$$

where $\mathrm{diag}\{\cdot\}$ is the diagonal operator and the vector inequalities are elementwise. Note that both $\boldsymbol{F}^{ext}$ and $T(x_t)$ are linear functions of the optimization variable $\boldsymbol{k}^d$, thus the last two inequality constraints can be easily expressed in the generic form $\boldsymbol{C}\boldsymbol{k}^d \leq \boldsymbol{d}$. Moreover, if $T(x_t) < \varepsilon$, then $\boldsymbol{k}^d = \boldsymbol{k}^{min}$. Note also that, since only the translational part of the impedance model is considered, the GMM is trained to encode only the linear part of velocities and forces.

8.3.3 Experiments and results
8.3.3.1 Experimental setup
The experiments conducted in this study encompass a twofold approach. Initially, human demonstrations are employed to facilitate the training of

the GMM. Subsequently, the framework's performance is assessed through the autonomous execution of a table cleaning task under two distinct conditions, which involve the task execution without external disturbances as well as with the introduction of external perturbations. This comprehensive set of experiments allows for the robust evaluation of the framework's capabilities and adaptability in practical scenarios. For each condition, three stiffness settings are tested:

1. Low constant stiffness (LS), where $k^d = k^{min}$;
2. High constant stiffness (HS), where $k^d = k^{max}$;
3. Optimized stiffness (OS), where k^d is found online through our QP formulation in (8.22).

The same desired end-effector trajectory generated by GMR is used for all the stiffness settings, while the desired interaction force is employed only by OS.

The experimental configuration for the human demonstrations is visually depicted in Fig. 8.9, concurrently illustrating the trajectory traversed by the human instructor. The human operator interfaces with the MOCA-MAN interface, utilizing the four buttons described in Section 8.3.2.1. The demonstrated trajectory commences from the "start" position, which corresponds to the upper-left corner of the table when viewed from the human's perspective. The human instructor subsequently guides the robot's end-effector along a path conducive to table cleaning, maintaining contact between the sponge and the table's surface while moving parallel to the shortest side of the table. Upon reaching the table's lower end, the human operator guides the robot through unconstrained motion to the upper region of the table, shifting it slightly to the right relative to the preceding trajectory by an extent equivalent to the width of the cleaning tool. This sequence is repeated a total of six times until the "goal" is achieved.

Throughout this process, three distinct demonstrations are conducted, all executed by the same human instructor. During these demonstrations, various data points, including the desired end-effector position, linear velocity, and interaction force with the environment, are meticulously recorded. These recorded data are subsequently harnessed for training the GMM, a crucial step in the learning process.

Throughout the demonstrations, a consistent Cartesian impedance profile is employed for the whole-body controller of MOCA. The impedance is set by a stiffness matrix $K^d = \text{diag}\{500, 500, 500, 50, 50, 50\}$ and a damping matrix computed utilizing a double diagonalization formula. Additionally, joint stiffness is introduced as a secondary task to avert the occurrence of joint limit violations and to maintain suitable manipulability configurations. Specifically, the joint stiffness and damping matrices K_0 and D_0, respectively, of size $\mathbb{R}^{n \times n}$, alongside the desired joint configuration q_0 in $\mathbb{R}^n$, work in concert to govern this secondary task.

Furthermore, the human instructor is endowed with the capacity to modulate the robot's admittance level across three discrete settings. These settings encompass low admittance, characterized by M^{adm} with diagonal values of $\{6, 6, 6\}$ and D^{adm} featuring diagonal elements of $\{40, 40, 40\}$; medium admittance, defined by M^{adm} with diagonal values of $\{4, 4, 4\}$ and D^{adm} with diagonal elements of $\{30, 30, 30\}$; and high admittance, with M^{adm} featuring diagonal values of $\{2, 2, 2\}$ and D^{adm} possessing diagonal elements of $\{20, 20, 20\}$. The transition between these admittance levels is facilitated by a dedicated button on the MOCA-MAN interface.

Moreover, the human instructor retains the ability to switch between two distinct operational modes: locomotion, characterized by substantial mobile base movements, and manipulation, in which the mobile base remains stationary. This transition is orchestrated through a button on the MOCA-MAN interface, with simultaneous adjustments made to H, K_0, and q_0. Notably, H and K_0 can adopt one of two predefined configurations. In manipulation mode, they assume the values $H = \text{diag}\{10 \cdot \mathbf{1}_{n_b}, 2 \cdot \mathbf{1}_{n_a}\}$ and $K_0 = \text{diag}\{2 \cdot \mathbf{1}_n\}$, while in locomotion mode, they are set to $H = \text{diag}\{2 \cdot \mathbf{1}_{n_b}, 10 \cdot \mathbf{1}_{n_a}\}$ and $K_0 = \text{diag}\{50 \cdot \mathbf{1}_n\}$. The choice of q_0 in this context aligns with the current arm configuration when the mode switch occurs. These versatile controls permit seamless adaptation to diverse operational requirements.

The identical experimental setup employed for the instructional demonstrations serves as the foundation for the autonomous task repetitions. These repetitions are subjected to scrutiny under two distinct conditions for each stiffness setting, specifically, LS, HS, and OS. In the first condition, denoted as autonomous task repetition without external disturbances, the robot replicates the previously instructed task within the unaltered environment featured during the instructional phase.

Conversely, in the second condition, referred to as autonomous task repetition with external disturbances, the environment encounters perturbations throughout the execution of the task. It is worth noting that the cleaning task entails six cleaning movements spanning from the upper region of the table to its lower end. These cleaning movements are interspersed with five instances of unconstrained free motions, allowing the robot to return to the uppermost section of the table before embarking on the subsequent cleaning movement. Several perturbations are introduced during the course of this experiment, further challenging the robot's performance:

1. Single slow table lifting and lowering during the first cleaning movement;
2. Single fast table lifting and lowering during the second cleaning movement;
3. Repeated low frequency and high amplitude table lifting and lowering during the third cleaning movement;
4. Repeated high frequency and low amplitude table lifting and lowering during the fourth cleaning movement;
5. Collision with a human during the fifth cleaning motion;
6. Collision with a human during the last free motion.

The values of the parameters used during the autonomous repetition experiments are $k^{min} = [200, 200, 200]$, $k^{max} = [1000, 1000, 1000]$, $F^{max} = [60, 60, 60]$, $\varepsilon = 0.4$, $x_t(0) = 1$, $Q = \text{diag}\{3200, 3200, 3200\}$, and $R = \text{diag}\{1, 1, 1\}$. The selection of the minimum stiffness values aimed at fostering high compliance is underpinned by the need to maintain acceptable position-tracking performance. In contrast, the determination of maximum stiffness values is driven by the aspiration to achieve precise and highly accurate motion tracking while upholding system stability. The establishment of a maximum force threshold aligns with the robot's payload capacity, ensuring that the applied forces remain within safe operational limits. Additionally, the initial energy level in the energy tank is deliberately set slightly above the specified energy threshold (ε) to safeguard against energy depletion.

Moreover, the experimental choices made regarding the matrices Q and R are motivated by the desire to accord higher priority to force-tracking objectives within the control framework, reflecting the significance of accurately tracking force-related parameters.

8.3.3.2 Experimental results

The results of the GMM training, the trajectories of the three demonstrations, and the output of the GMR for the desired position are reported in Fig. 8.10, where the GMR output well approximates the three demonstrations.

The outcomes of the autonomous task repetition experiments conducted without external disturbances, encompassing LS, HS, and OS, are

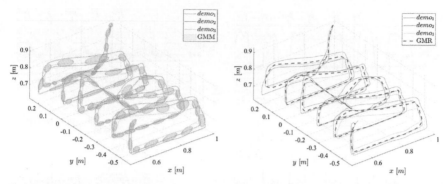

Figure 8.10 The position learning results of GMM (left)/GMR (right). Three demonstrations were used.

graphically depicted in Fig. 8.11. The evaluations encompass an assessment of both force tracking and position tracking performance, while specific details pertaining to stiffness values and energy levels within the tank are exclusively provided for the OS condition.

In the case of LS (Fig. 8.11(b)), the behavior exhibits commendable compliance but fails to accurately track the desired force component along the z axis, which is vital for the effective execution of the cleaning task. On the contrary, HS (Fig. 8.11(c)) manifests an excessive force exertion along the z axis, leading to a stick-and-slip phenomenon. This phenomenon is evident from the observed trends in both force and position along the x axis.

In contrast, the OS condition (Fig. 8.11(a)) is characterized by the generation of an appropriate level of compliance essential for faithfully tracking the desired force during interactions with the table. Simultaneously, it maintains a low stiffness during periods of free motion, where no interactions are required. The dynamics of energy within the tank are shaped by two primary components: the dissipated energy and the exchanged energy due to fluctuations in stiffness, as per Eq. (8.20). Despite occasional energy drops attributed to stiffness variations, the energy level within the tank exhibits an overall upward trend, never falling below the designated threshold ε. This consistently positive trend underscores the preservation of system passivity, validating the stability of the control framework.

The outcomes of the autonomous task repetition experiments conducted under the influence of external disturbances are depicted in Fig. 8.12, where we evaluate the performance of LS, HS, and OS.

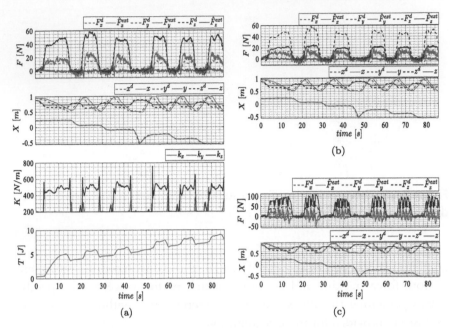

Figure 8.11 Results of autonomous task execution experiments in nominal conditions: (a) OS settings, from top to bottom are desired interactive force F^d and estimated force $\hat{F}^{ext}$ during experiments, desired position X^d and real position X, generated stiffness K, and tank energy $T(x_t)$; (b) LS settings, from top to bottom are desired interactive force F^d and estimated force $\hat{F}^{ext}$ during experiments, desired position X^d and real position X; and (c) HS settings, same contents with (b).

In Fig. 8.12(b), we observe the results for LS. While the robot exhibits consistent compliance to the environmental uncertainties introduced by external disturbances, it faces challenges in performing the task due to an insufficient force applied to the table. In stark contrast, the HS condition (Fig. 8.12(c)) reveals that, as soon as the table is lifted, the interaction force surpasses the robot's payload limit. A similar issue arises when a human collides along the x axis, as illustrated in the right portion of Fig. 8.12(c).

Fig. 8.12(a) presents the results for OS, which demonstrates a notable ability to adapt to various types of disturbances applied along the z axis. In the presence of high-frequency disturbances, the force tracking task becomes more challenging, albeit still yielding acceptable performance. Along the x axis, OS maintains low stiffness, resulting in a compliant response

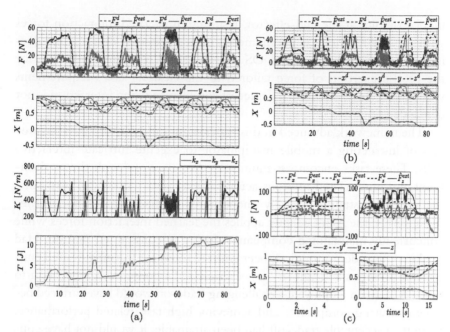

Figure 8.12 Results of the autonomous task execution experiments with unplanned disturbances: (a) OS; (b) LS; and (c) HS. The plotted variables are similar to those in Fig. 8.11. Only the (c) figure showing disturbances 1) and 5) is different since we could not report the whole task execution due to repetitive stop of the robot due to a high acceleration exceeding the safety limits.

when the robot interacts with the table or encounters human disturbances in that direction during both interaction and free motion. Furthermore, the energy within the tank consistently remains above the specified threshold, exhibiting an overall increasing trend. These observations highlight how external disturbances influence the additional energy injected (e.g., during table lifting) and extracted (e.g., during table lowering) from the tank.

8.3.3.3 Discussion and conclusion

The findings reported in Section 8.3.3.2 demonstrate that, on the whole, the OS approach surpasses both the LS and HS alternatives. While LS effectively maintains a high degree of compliance and resilience to external perturbations and uncertainties, it falls short of achieving satisfactory performance in the cleaning task. In contrast, HS exhibits notable deficiencies in performance, safety, and robustness, primarily due to its propensity to induce rigidity at the end-effector, resulting in elevated interaction forces with the environment as it vehemently rejects external disturbances during motion. On the contrary, OS adeptly strikes a balance, allowing for the precise exertion of force tailored to the cleaning task's requirements while preserving a substantial degree of compliance when interaction is not demanded.

The framework advanced in this paper significantly streamlines the process of instructing a mobile manipulator to execute intricate interactive tasks. These tasks encompass multifaceted interactions with the environment and unrestricted motions across an extensive workspace, surpassing the spatial constraints typically encountered by fixed-base robotic arms. Additionally, the robot adeptly applies the learned interaction forces while concurrently maintaining compliance during unconstrained motion. Had the kinesthetic demonstrations been utilized solely for learning desired end-effector positions and velocity trajectories, and a constant stiffness, akin to HS and LS, been employed, a challenging trade-off would have surfaced between ensuring compliance and achieving high task-related performance. Even if an acceptable trade-off had been attainable, it would not have conferred the same robustness against external disturbances that our framework systematically achieves. Furthermore, our approach systematically preserves the system's passivity via an energy tank-based constraint and accommodates payload limits and stiffness bounds as integral constraints within the optimization problem, bolstering the system's safety and adaptability in practical applications.

Future research endeavors will encompass an expanded scope of the learning process, aiming to encompass all pertinent parameters associated with the Cartesian impedance whole-body controller. These parameters include the comprehensive array of motion modes and secondary tasks, which will be mastered through the MOCA-MAN interface. The current study adopted a task that allowed for the a priori selection of specific controller parameters, ensuring the proficient execution of the robotic platform's loco-manipulation behavior throughout the task. However, in more intricate and dynamically evolving environments, such preselection is not always tenable, necessitating more adaptive approaches. Furthermore, a compelling avenue for future exploration pertains to the autonomous self-tuning of the weighting matrices within the QP formulation, denoted as $\mathbf{Q}$ and $\mathbf{R}$. This development holds promise for enhancing the controller's adaptability and performance, particularly in scenarios marked by varying dynamics and operational conditions.

8.4 Chapter summary

Due to the manufacturing shift from mass to customized production processes and the trend of an aging population in recent years, collaborative robots (fixed- and mobile-base) have been created to solve this problem. Meanwhile, such robots come out of the cage and can work with human partners. In this chapter, we showed how the role allocation algorithm can be applied in a hierarchical and sequential task. Besides, with the help of the MOCA-MAN interface, we showed humans and CMMs can work together efficiently in a comanipulation manner by combining each other's strengths. Furthermore, humans can not only be the robots' partners but also teachers for the CMMs through the MOCA-MAN interface. Specifically, the proposed offline learning and online real-time optimization framework make CMMs interact with a highly dynamic environment safely, which is quite a challenging task.

We believe that with the properly designed interface and the artificial intelligence techniques, better physical human–robot interaction and collaboration can be achieved with suitable operation modes between each other.

References

Abu-Dakka, F.J., Rozo, L., Caldwell, D.G., 2018. Force-based variable impedance learning for robotic manipulation. Robotics and Autonomous Systems 109, 156–167.

Abu-Dakka, F.J., Saveriano, M., 2020. Variable impedance control and learning—a review. Frontiers in Robotics and AI, 177.

Ajoudani, A., Zanchettin, A.M., Ivaldi, S., Albu-Schäffer, A., Kosuge, K., Khatib, O., 2018. Progress and prospects of the human–robot collaboration. Autonomous Robots 42, 957–975.

Averta, G., Hogan, N., 2020. Enhancing robot–environment physical interaction via optimal impedance profiles. In: 2020 8th IEEE RAS/EMBS International Conference for Biomedical Robotics and Biomechatronics (BioRob), pp. 973–980.

Balatti, P., Fusaro, F., Villa, N., Lamon, E., Ajoudani, A., 2020a. A collaborative robotic approach to autonomous pallet jack transportation and positioning. IEEE Access 8. https://doi.org/10.1109/ACCESS.2020.3013382.

Balatti, P., Kanoulas, D., Tsagarakis, N., Ajoudani, A., 2020b. A method for autonomous robotic manipulation through exploratory interactions with uncertain environments. Autonomous Robots 44, 1395–1410.

Bishop, C.M., Nasrabadi, N.M., 2006. Pattern Recognition and Machine Learning, vol. 4. Springer.

Buchli, J., Stulp, F., Theodorou, E., Schaal, S., 2011. Learning variable impedance control. The International Journal of Robotics Research 30, 820–833.

Calinon, S., Lee, D., 2019. Learning control. In: Humanoid Robotics: A Reference. Springer, pp. 1261–1312.

Calinon, S., Sardellitti, I., Caldwell, D.G., 2010. Learning-based control strategy for safe human–robot interaction exploiting task and robot redundancies. In: 2010 IEEE/RSJ International Conference on Intelligent Robots and Systems. IEEE, pp. 249–254.

Chang, C., Slagle, J., 1971. An admissible and optimal algorithm for searching and/or graphs. Artificial Intelligence 2, 117–128.

Duan, J., Gan, Y., Chen, M., Dai, X., 2018. Adaptive variable impedance control for dynamic contact force tracking in uncertain environment. Robotics and Autonomous Systems 102, 54–65.

Ellekilde, L.P., Christensen, H.I., 2009. Control of mobile manipulator using the dynamical systems approach. In: 2009 IEEE International Conference on Robotics and Automation. IEEE, pp. 1370–1376.

El Makrini, I., Omidi, M., Fusaro, F., Lamon, E., Ajoudani, A., Vandcrborght, B., 2022. A hierarchical finite-state machine-based task allocation framework for human–robot collaborative assembly tasks. In: 2022 IEEE/RSJ International Conference on Intelligent Robots and Systems (IROS), pp. 10238–10244.

Ferraguti, F., Preda, N., Manurung, A., Bonfe, M., Lambercy, O., Gassert, R., Muradore, R., Fiorini, P., Secchi, C., 2015. An energy tank-based interactive control architecture for autonomous and teleoperated robotic surgery. IEEE Transactions on Robotics 31, 1073–1088.

Ficuciello, F., Villani, L., Siciliano, B., 2015. Variable impedance control of redundant manipulators for intuitive human–robot physical interaction. IEEE Transactions on Robotics 31. https://doi.org/10.1109/TRO.2015.2430053.

Fusaro, F., Lamon, E., De Momi, E., Ajoudani, A., 2021a. An integrated dynamic method for allocating roles and planning tasks for mixed human–robot teams. 2021 30th IEEE International Conference on Robot & Human Interactive Communication (RO-MAN), 534–539.

Fusaro, F., Lamon, E., De Momi, E., Ajoudani, A., 2021b. A human-aware method to plan complex cooperative and autonomous tasks using behavior trees. https://doi.org/10.1109/HUMANOIDS47582.2021.9555683.

Gandarias, J., Balatti, P., Lamon, E., Lorenzini, M., Ajoudani, A., 2022. Enhancing flexibility and adaptability in conjoined human–robot industrial tasks with a minimalist physical interface. https://doi.org/10.1109/ICRA46639.2022.9812225.

Hart, P.E., Nilsson, N.J., Raphael, B., 1968. A formal basis for the heuristic determination of minimum cost paths. IEEE Transactions on Systems Science and Cybernetics 4, 100–107.

Hjorth, S., Lamon, E., Chrysostomou, F., Ajoudani, A., 2023. Design of an energy-aware Cartesian impedance controller for collaborative disassembly. In: 2023 International Conference on Robotics and Automation (ICRA).

Huang, Y., Rozo, L., Silvério, J., Caldwell, D.G., 2019. Kernelized movement primitives. The International Journal of Robotics Research 38, 833–852.

Ijspeert, A.J., Nakanishi, J., Hoffmann, H., Pastor, P., Schaal, S., 2013. Dynamical movement primitives: learning attractor models for motor behaviors. Neural Computation 25, 328–373.

Kim, W., Balatti, P., Lamon, E., Ajoudani, A., 2020. MOCA-MAN: A mobile and reconfigurable collaborative robot assistant for conjoined human–robot actions. https://doi.org/10.1109/ICRA40945.2020.9197115.

Lamon, E., De Franco, A., Peternel, L., Ajoudani, A., 2019. A capability-aware role allocation approach to industrial assembly tasks. IEEE Robotics and Automation Letters 4, 3378–3385.

Lamon, E., Fusaro, F., Balatti, P., Kim, W., Ajoudani, A., 2020a. A visuo-haptic guidance interface for mobile collaborative robotic assistant (MOCA). In: 2020 IEEE/RSJ International Conference on Intelligent Robots and Systems (IROS), pp. 11253–11260.

Lamon, E., Leonori, M., Kim, W., Ajoudani, A., 2020b. Towards an intelligent collaborative robotic system for mixed case palletizing. https://doi.org/10.1109/ICRA40945.2020.9196850.

Lamon, E., Peternel, L., Ajoudani, A., 2018. Towards a prolonged productivity in Industry 4.0: A framework for fatigue minimisation in robot–robot co-manipulation. https://doi.org/10.1109/HUMANOIDS.2018.8625051.

Martelli, A., Montanari, U., 1973. Additive AND/OR graphs. In: Proc. 3rd International Joint Conference on Artificial Intelligence, pp. 1–11.

Merlo, E., Lamon, E., Fusaro, F., Lorenzini, M., Carfi, A., Mastrogiovanni, F., Ajoudani, A., 2022. Dynamic human–robot role allocation based on human ergonomics risk prediction and robot actions adaptation. In: 2022 International Conference on Robotics and Automation (ICRA), pp. 2825–2831.

Merlo, E., Lamon, E., Fusaro, F., Lorenzini, M., Carfi, A., Mastrogiovanni, F., Ajoudani, A., 2023. An ergonomic role allocation framework for dynamic human–robot collaborative tasks. Journal of Manufacturing Systems 67. https://doi.org/10.1016/j.jmsy.2022.12.011.

Michel, Y., Rahal, R., Pacchierotti, C., Giordano, P.R., Lee, D., 2021. Bilateral teleoperation with adaptive impedance control for contact tasks. IEEE Robotics and Automation Letters 6, 5429–5436.

Ott, C., 2008. Cartesian Impedance Control of Redundant and Flexible-Joint Robots. Springer.

Peternel, L., Petrič, T., Oztop, E., Babič, J., 2014. Teaching robots to cooperate with humans in dynamic manipulation tasks based on multi-modal human-in-the-loop approach. Autonomous Robots 36, 123–136.

Wu, Y., Balatti, P., Lorenzini, M., Zhao, F., Kim, W., Ajoudani, A., 2019. A teleoperation interface for loco-manipulation control of mobile collaborative robotic assistant. IEEE Robotics and Automation Letters 4, 3593–3600.

Wu, Y., Lamon, E., Zhao, F., Kim, W., Ajoudani, A., 2021. Unified approach for hybrid motion control of MOCA based on weighted whole-body Cartesian impedance formulation. IEEE Robotics and Automation Letters 6, 3505–3512.

Wu, Y., Zhao, F., Tao, T., Ajoudani, A., 2020. A framework for autonomous impedance regulation of robots based on imitation learning and optimal control. IEEE Robotics and Automation Letters 6, 127–134. https://doi.org/10.1109/lra.2020.3033260.

Zhang, L., Zhao, J., Lamon, E., Wang, Y., Hong, X., 2023. Energy efficient multi-robot task allocation constrained by time window and precedence. IEEE Transactions on Automation Science and Engineering.

Zhao, J., Giammarino, A., Lamon, E., Gandarias, J.M., De Momi, E., Ajoudani, A., 2022. A hybrid learning and optimization framework to achieve physically interactive tasks with mobile manipulators. IEEE Robotics and Automation Letters 7, 8036–8043.

CHAPTER 9

Case studies of proactive human–robot collaboration in manufacturing

In accordance with the human-centric, sustainable, and resilient principles outlined in Industry 5.0, Proactive HRC presents wide-ranging applications across various industries, including machining, assembly, welding, and more. Proactive HRC establishes a collaborative system, allowing robots to execute high-load and high-risk operations, while humans perform ultra-flexible tasks. Through this approach, Proactive HRC optimizes processing steps, enhances information systems, and refines human manual labor. This chapter presents two case studies of Proactive HRC in manufacturing. The first case focuses on the disassembly of aging electric vehicle batteries. We provide relevant algorithms and system demonstration of disassembly work as a reference for those interested in this field. In the context of human–robot collaborative drilling, research is still in the exploratory stage with preliminary results. We propose a conceptual system and discuss the enabling technologies to elaborate our insights and future expectations for the potential collaborative drilling system for aircraft skin.

9.1 Disassembly of aging electric vehicle batteries

In recent years, there has been a global push for green manufacturing and sustainable development. Electric Vehicles (EVs), as representatives of clean energy, have experienced rapid development and an increasing electrification ratio. However, with the widespread adoption of EVs and the prolonged service life, the issue of handling aging EVB has become increasingly prominent. Recycling important materials such as cobalt, nickel, and lithium in aging batteries can reduce mining and environmental pollution while also helping to minimize resource waste. Disassembly and reuse are currently the most effective processing methods. However, due to the diverse designs of EVB across vehicle models and the varied historical service conditions of each aging EVB, achieving batch operations for disassembly aging EVB challenging, requiring a high level of flexibility and complexity.

Proactive Human–Robot Collaboration Toward Human-Centric Smart Manufacturing
https://doi.org/10.1016/B978-0-44-313943-7.00016-8

229

In order to improve the efficiency of disassembly and assembly during the remanufacturing cycle and promote the industrialization of intelligent disassembly and assembly operations for retired products, the manufacturing mode of collaborative work between human operators and collaborative robots has received unprecedented attention and widespread application. In this section, a study on collaborative disassembly tasks between humans and robots is conducted using three types of retired batteries as examples. The objectives are as follows:

a) to construct a scene model and task representation mode that is suitable for human–robot collaborative disassembly;

b) to generate human–robot collaboration strategies using reinforcement learning methods and transfer learning methods for similar tasks;

c) to demonstrate the feasibility of the intelligent strategy generation method through actual disassembly results.

9.1.1 Disassembly environment

9.1.1.1 Collaborative work scene

The human–robot collaborative work scene aims to achieve information and operational interactions between physical entities, with the goal of jointly completing tasks through mutual collaboration, as shown in Fig. 9.1. The aging EVB disassembly environment for human–robot collaboration is illustrated in the figure, consisting of four components: the physical entity elements and interaction control elements form the physical work scene, while the digital work scene is composed of computational elements and information model elements. Among them, the interaction control elements serve as the mediators between the physical and digital scenes, with their digitized information used to drive the physical model and serve as the iterative foundation for the information elements.

The physical entity elements of the human–robot collaborative disassembly operation scenario mainly consist of the operator (operating worker, collaborative robot) and the operating object (task object, operating tool).

In the operator module, the operating worker has high flexibility and strong decision-making ability, which enables them to easily handle the disassembly and assembly operations of flexible objects. However, humans are prone to fatigue, and prolonged work leads to decreased willingness to perform tasks, poor task execution stability, and a certain probability of making mistakes. On the other hand, robots have the opposite characteristics. They can ensure high-precision repetitive actions but have lower adaptability to different tasks. Therefore, the operating module formed by

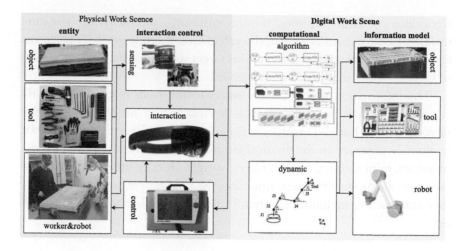

Figure 9.1 Human–robot collaborative work scene.

the complementary advantages of humans and robots can highly adapt to various disassembly and assembly tasks. The operating object module includes the various components to be operated and various operating tools used in task execution. The components of the product are classified into main parts and fasteners and are placed in fixed areas on the operating table. In the actual operation process, it is usually agreed that the operating tools follow the principle of being placed in a fixed position for easy access and immediate return after use, in order to improve the efficiency of tool retrieval.

In the human–robot integrated operation scenario, the operator usually only needs to acquire information and further understand the current environmental status based on the acquired information in order to take appropriate actions. Humans obtain and interact with information through direct observation or visualized interactive control devices, while robots usually require additional auxiliary devices for information acquisition from the external environment. Therefore, the interactive control elements defining the integration scenario usually consist of sensing devices, control devices, and interaction devices.

As the lifeblood of intelligent manufacturing processes, data is crucial for intelligent perception, interaction, and decision-making behaviors. In the integration system, the decision-making and planning module is mainly used to determine the next action strategy of the robot based on the information obtained by the perception module. This module usually includes algorithms for path planning, motion planning, and behavior decision-making. In the dynamics model, the calculation module typically performs kinematic calculations based on various target states to generate a collection of variable values used to drive the model.

The information model usually refers to the twin mapping of physical entities, which not only serves as a visualization model mapping the physical entities but also includes various attribute information that cannot be directly observed in the physical entities, such as three-dimensional dimensions, joint torques, contact forces, weight, density, and state attributes. Based on the product's information data, the task information model is further constructed, including constraint models, complexity models, and so on.

9.1.1.2 Disassemble task analysis

The research focuses on the disassembly process of EVB-C, as shown in Fig. 9.2, the conventional disassembly process requirements are as follows:

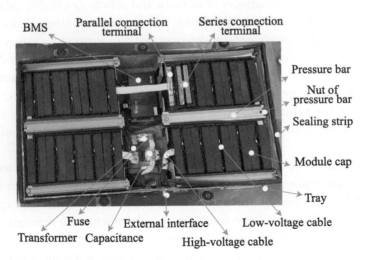

Figure 9.2 Physical diagram of the internal structure of the EVB-C pack.

a) (Cover removal) This includes removing the screws connecting the cover and the tray, taking out the screws and placing them in the designated location for sorting. Then, switch the suction cup end effector, lift the top cover, and place it in the specified position. The sealing strip at the connection between the top cover and the tray should also be removed.

b)   (Pressure bar removal) Firstly, switch the end effector tool and remove the nuts fixing the pressure bar. Some nuts may have excessive pretightening force, which cannot be directly operated by the robot. In such cases where the robot cannot successfully execute the task, it is handed over to humans.

c)   (High-voltage cable removal) Firstly, remove the insulation protective cover at the top of the module, check the remaining voltage of the battery, remove the screws fixing the high-voltage cables, and detach the connecting constraints. Interact with humans to remove the cables.

d)   (Low-voltage cable removal) Continue removing the screws that fix the low-voltage cables at the top of the battery module, and release the connection between the low-voltage harness and the battery module.

e)   (Module removal) Further remove the series terminals and fixing screws between modules, separate each module, and take them out in order.

f)   (Cable cleaning) Remove the remaining cables, disconnect the plugs between cables and the controller, take out the cables from the buckle. For cables fixed with zip ties, the ties need to be cut first.

g)   (Other parts) Sequentially remove the remaining insulation shock pads and safety control systems, such as fuse boxes.

9.1.2 Experimental method

The human–robot collaborative work environment is based on the UR5 collaborative robot and the ROS as the software and hardware platform. The robot end effector is equipped with a two-finger gripper, suction cup, electric screwdriver, and other tools. Furthermore, the human–robot collaborative work platform integrates the key technical methods described earlier and incorporates safety contact force limiting capabilities to ensure robot contact safety. Reinforcement learning and transfer learning empower the robot with the ability to learn and reuse experiences, while also possessing the capability to make task judgments and spatial localization based on visual information, enabling accurate execution of disassembly and assembly operations and collaborative behaviors. Visualized interaction provides operational guidance to ensure operational safety.

Based on the aforementioned technologies and methods, a human–robot collaborative simulation system is developed using programming in Python, Solidworks as the modeling software, robot simulation development software ROS, 3D rendering engine Unity, augmented reality toolkit ARToolKit, depth camera RealSense, and Microsoft's augmented reality glasses HoloLens. The system is validated using the disassembly of aging EVB as an application case.

9.1.2.1 Task representation

The disassembly and recycling of aging EVB packs typically involve four main steps: structural modular disassembly, cable removal, cell disassembly, and cell recycling. Due to considerations of the stability and safety of the battery cells in aging EVB, this experiment mainly focuses on the disassembly of structural modules and cables, stopping at the battery module without further disassembling the cells.

Since traditional disassembly or assembly tasks are entirely carried out by manual labor, it assumes that humans have the ability to complete all tasks independently. Therefore, for operators, their range of motion can include the selection of all components. In contrast, robots are relatively less flexible and may have limited capabilities when dealing with nonrigid hoses and cable connections. In such cases, manual intervention or collaboration with humans may be required.

General EVB packs consist of various electronic components, including battery trays, battery modules, insulation damping pads, sealing strip, cables, Battery Management System (BMS) modules, and fuse modules. The cables include high- and low-voltage cables. The general processing flow for aging EVB packs is shown in Fig. 9.3, which mainly involves three stages: battery pack disassembly, module disassembly and recycling, and module recombination for reuse.

When disassembly aging EVB packs, besides various constraints, the general disassembly tasks need to follow the following basic principles. The disassembly process should follow a reasonable operation sequence, usually starting with the removal of external accessories, followed by internal component removal, and finally disassembling the components into parts. Taking the disassembly process of the battery pack as an example, the first step is to remove the external protective cover, then remove various cables and electronic components, and finally dismantle the modules.

For the expression of disassembly tasks, existing methods based on logic gates or graphs have achieved a unified and standardized representation of assembly/disassembly tasks in human–robot collaborative scenarios. However, these representation methods only include the relationships and constraints between parts, with weak expressive power for products with a large number of components and complex operational processes. They also lack the associated information between task operations and the characteristics

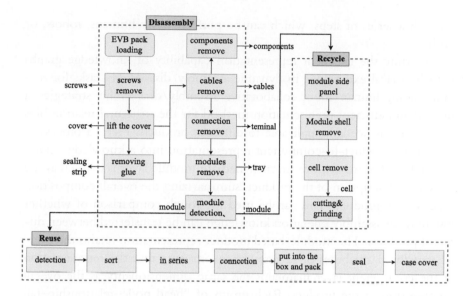

Figure 9.3 General EVB pack handling process.

of human–robot collaborative work. In contrast to the representation form of a structure tree, knowledge graph can store rich attribute information for each node and edge, and its representation based on triplets is more convenient for computation and search. Therefore, this section proposes a knowledge graph representation method for human–robot collaborative tasks.

The multidimensional data involved in the human–robot collaborative assembly/disassembly process, including products, equipment (industrial robots, operating tools), and processes, are treated as the data layer. The product pattern layer, assembly/disassembly task pattern layer, and operation process pattern layer are constructed separately. By combining the data layer with the pattern layers, the knowledge-based description of multidimensional information in the human–robot collaborative assembly/disassembly process is achieved. Specifically, the assembly/disassembly product pattern layer mainly describes the structural features of the product and the division of its components. The assembly/disassembly task pattern layer mainly describes the multiple task units defined for completing the assembly/disassembly operation, reflecting only the procedural goals of the operation. The assembly/disassembly operation process pattern layer mainly describes the specific execution process of the assembly/disassembly task, i.e., how to achieve the assembly/disassembly of product components through a series of steps, which can be performed by humans, robots, or both.

To study the excellent representation capability of knowledge graphs with historical experience in dynamic assembly/disassembly tasks for recommending human–robot collaborative assembly/disassembly strategies, a unified and efficient representation method for the structural relationships of products is needed, applicable to different products. Here, a process of converting product-to-component representation into a knowledge graph is established. Furthermore, an intermediate module layer is defined as the structural description of the product, summarizing the overall composition types of the product. This can be used for rough comparison of whether assembly/disassembly operation knowledge can be transferred between different products. The module-based product representation retains the overall structured attributes of subassemblies, ensuring that the representation at the component layer maintains the same dimensionality as the overall composition of the product. Each group of "head node–relationship–tail node" and its associated attribute information in the knowledge graph at the component layer can be used for detailed strategy recommendation and discrimination. The process of transforming models into knowledge graphs is illustrated in Fig. 9.4(a) represents the mapping relationship between product component information and graph nodes, while the knowledge representation for complete products is shown in Fig. 9.4(b).

As shown in Fig. 9.5, the product ontology layer for disassembly tasks includes six types of entities: product entity, subassembly entity, component unit entity, fastener entity, part entity, and corresponding operating tools. The product entity contains attributes such as name, batch model, size standard, and functional type. The subassembly and component unit entities only include name and function attributes. The smallest units that constitute the product are the part entity and fastener entity, both of which contain attributes such as name, universal index ID, operating process characteristics, size model, and material. As for operating tools, they are usually named based on their own functional attributes and serve as the name attribute, while the model attribute includes the detailed classification and size of the tool. As external nodes of the product, operating tools have "Used to" and "Need to" relationships with the objects they operate on, indicating whether components require the use of tools for operation. There is a "Has Subassembly" relationship between the product and subassembly entities, a "Has Component" relationship between the component unit and product entities, and the inclusion of parts in both

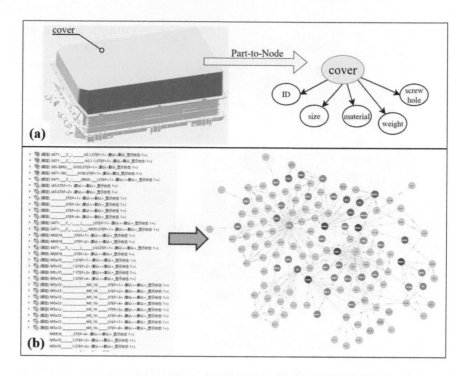

Figure 9.4 Product structure representation based on knowledge graph.

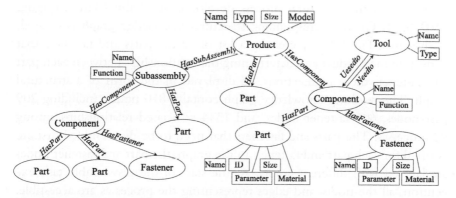

Figure 9.5 Product ontology layer for disassembly tasks.

component units and products is represented by the "Has Part" relationship. The component unit and fastener entities have a "Has Fastener" relationship.

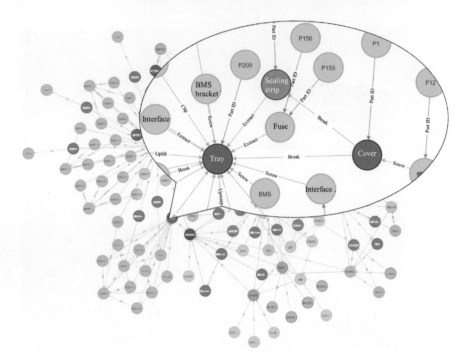

Figure 9.6 Knowledge graph representation of the EVB-C.

Based on the structural process development of the C–model aging EVB (EVB-C), its corresponding structural knowledge graph is created, as shown in Fig. 9.6. By creating a list of all the parts and fasteners that constitute the product and determining the relationships between each part and fastener, these connections are displayed in the form of a structural graph. The resulting knowledge graph contains 810 nodes, including 209 part nodes, 601 fastener nodes, and 1344 associated relationships among 22 processes. The parts and fasteners that need to be disassembled first are verified for each part and fastener. In the graph, the part and fastener nodes are labeled with different colors. At the start of the disassembly task execution, all the nodes and edges representing the processes are accessible, and the arrows between nodes in the graph indicate the sequence of individual fasteners and parts. For example, before removing the "top cover" (P1), the 19 screws connecting it to the battery tray (P209) must be loosened.

During the initialization stage of the task, all nodes in the structural knowledge graph of the complete product are in a connected state. As the

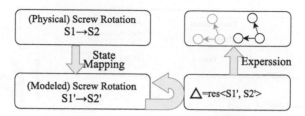

Figure 9.7 Disassembly status update based on dynamic graph representation.

task progresses, detached parts are detected one by one. Based on model matching, the corresponding nodes for these parts are identified, and the corresponding nodes are released based on the removal relationship command, indicating the completion of disassembly for that part. This process is illustrated in Fig. 9.7.

9.1.2.2 Collaborative strategy generation

1. Collaborative strategy self-optimization architecture

For intelligent agents performing human–robot collaborative disassembly tasks, simply setting constraints such as time optimization or minimum difficulty for decision planning often results in a task division where each component is executed by the most capable operator, without considering the overall efficiency of collaborative work and the problem of decreasing human operational stability during a series of execution processes. By using reinforcement learning, the dynamic operation process is considered based on the operator's state, task difficulty, and corresponding execution time, and through rewards, operators and robots continuously try and learn to obtain an efficient task allocation mode.

The self-reinforcement learning architecture for human–robot collaboration strategy based on reinforcement learning is shown in Fig. 9.8, where the task environment includes product components to be operated on and related operating tools, and the action spaces a_H and a_R are set for human and robot agents, respectively. The task execution mode division between humans and robots, or their collaboration, is determined by adjusting the task operation sequence between them in the reinforcement learning environment. In the decision-making process of assigning new tasks, corresponding reward functions are set according to changes in task status. The critic network is used to evaluate the decision of the actor network.

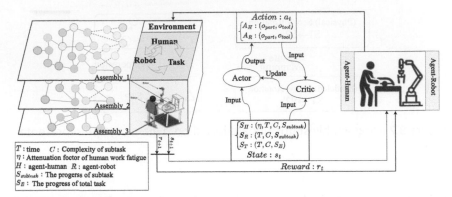

Figure 9.8 A human–robot collaborative decision-making architecture based on reinforcement learning.

2. Human–robot collaborative reinforcement learning network

The Human–Robot Collaborative Reinforcement Learning (HRC-RL) framework is designed by extending the Deep Deterministic Policy Gradient (DDPG). As illustrated in Fig. 9.9, the action networks of the robot agent (Agent-Robot) and the human agent (Agent-Human) process the current state and generate their respective actions (A_R, A_H). Subsequently, the evaluation network assesses and computes the action values Q_R and Q_H. Meanwhile, due to the impact of the actions taken by both agents on the environment, the current state S transitions to the next state S'. Then, within the dual-agent cooperative reinforcement learning network, the same Actor–Critic network is utilized to solve the value functions $Q_{R'}$, and $Q_{H'}$, corresponding to the next state S', and the TD-error from the same evaluation network directly updates the corresponding parameters. Additionally, the TD-error between Q_{global} and Q'_{global}, indirectly updates Critic1 and Critic2. TD-error3, by incorporating global information into each evaluation network, indirectly optimizes the Actor network, effectively preventing individual agents from converging to local optima and improving exploration efficiency.

Compared with Multi-Agent Deep Deterministic Policy Gradient (MADDPG), the two agents in the HRC-RL framework have different execution capabilities. The action network is not only influenced by the environmental state, but also constrained by the human's own behavioral ability. In addition, noise is added to the local operator network of Agent-Human to maximize the robustness of the system.

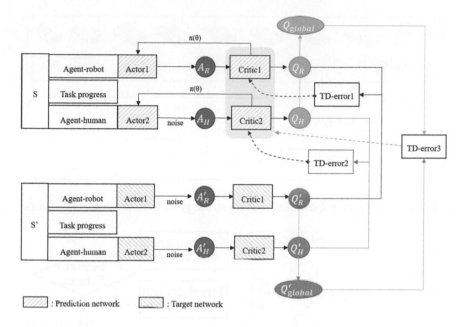

Figure 9.9 Human–robot collaborative reinforcement learning network.

3. Process adaptive method based on transfer learning

For iterative products of different models or batches of the same type, their overall structures are similar and the differences in the overall assembly structure are relatively small. Taking two models of aging EVB for a certain electric vehicle as an example, as shown in Fig. 9.10, the similarity of their product compositions is high, but they have different numbers of components and structural sizes, which leads to inconsistent task sequences. If we ignore the similarity between different models of products and perform completely new task planning for each model, it will significantly increase the iteration cost of the products. Therefore, based on transfer learning, by adaptively adjusting the domain of similar task policies, a collaborative strategy that meets the current task can be quickly generated, improving decision efficiency and reducing the re-planning cost of similar tasks.

4. Transfer learning model

To further achieve cross–domain policy transfer, a domain-adversarial discriminative adaptation network model is proposed to discriminate the correlation between the source domain task and the target domain task features in the workspace, and adjust the network parameters of the encoder in the target domain accordingly.

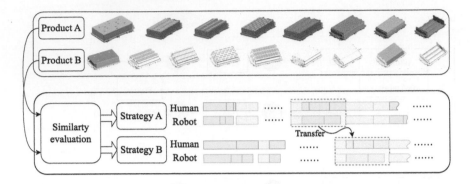

Figure 9.10 Human–robot collaboration strategy transfer for similar products.

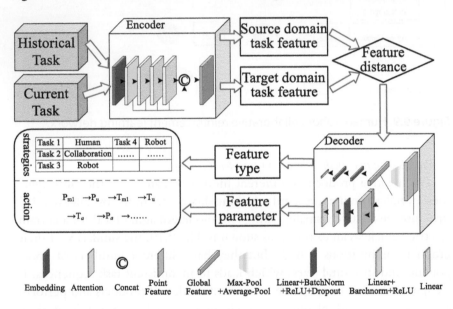

Figure 9.11 Strategy transfer architecture for different tasks.

For historical tasks that meet the transfer requirements in terms of similarity values and the current task, the human–robot collaboration strategies of the historical tasks are first extracted and mapped to the task states. This enables the further generation of collaborative operation strategies for the target domain task based on domain adaptation. As shown in Fig. 9.11, the encoder network is shared between the historical disassembly tasks and the current disassembly task to generate task features for both the source domain and the target domain. In the encoder, point features are extracted using a joint model of linear and regularization methods. Additionally, in the discriminator network, the domain category of the feature for the upcoming operation task is determined. After adversarial training, the target encoder will have a similar feature distribution as the source domain. Therefore, the source domain decoder can be directly used for feature classification and segmentation. Next, through feature distance discrimination, similar features between two tasks are identified, and the shared features are decoded by the decoder to obtain the subtask types and their corresponding collaborative strategy action parameters.

5. Integrated collaboration based on transfer reinforcement hybrid drive

The decision-making methods of robots include various approaches, such as rule-based methods, neural network-based methods, and evolutionary algorithm-based methods. Imitation learning, transfer learning, and reinforcement learning exhibit different advantages. Among them, the training process of imitation learning requires a large amount of real experimental data to support, while transfer learning-based methods typically require a source domain or metamodel applicable to the current task type for reuse, also requiring a certain amount of historical data, but the amount of data and the difficulty of obtaining data required for transfer learning are much lower than that of imitation learning. In addition, as shown in Chapter 7, using the self-decision-making method of intelligent agent reinforcement learning allows robots to learn new strategy experience in different tasks. However, in the process of reinforcement learning, different types of tasks require different objective functions and reward functions, and the setting of these functions often depends on prior knowledge and experience of specific tasks. Therefore, for robots, the process of learning and exploration from scratch requires a lot of time for trial and error and learning.

In order to effectively utilize historical data and experience in the process of reinforcement learning, improve the efficiency and accuracy of robot learning, and fill the problem of insufficient source domain data in transfer learning-based methods, here proposes a transfer reinforcement mixed-driven decision adaptive method, as shown in Fig. 9.12. The proposed method is an implementation of a reinforcement learning method based on similar task policy transfer guidance, which can achieve the reuse of similar task policies and adaptively adjust according to different tasks and assembly features. This method improves the collaboration ability and autonomous operation ability of robots and human participants through more efficient

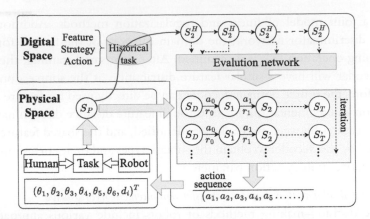

Figure 9.12 Intelligent decision framework based on transfer and reinforcement.

decision-making methods, thereby improving the efficiency of the entire assembly process.

a) (Reinforcement Learning Self-Decision) The reinforcement learning self-decision module primarily takes the current operational scene state as input, including the execution status of the task and the behavioral state of the human and robot. It then interacts with the intelligent agent and the operational space in the reinforcement learning space to determine the decision behavior suitable for the current state.

b) (Transfer Learning Decision Reuse) The transfer learning decision reuse module serves as an experience pool for historical decision tasks. Its main function is to provide policy transfer for similar tasks, reducing the cost of redecision for similar tasks. In the task transferability evaluation network, the similarity discrimination results between the source domain task and the target domain task serve as the criteria for transferability discrimination.

c) (Transfer Reinforcement Adaptation) For cases where there are differences between the source domain and the target domain that cannot be fully transferred, a transfer reinforcement mixed-driven adaptive decision method is employed. The tasks are divided, and policy transfer is conducted for similar subtasks. For tasks with lower similarity, reinforcement learning methods are used for decision supplementation to ensure adaptive decision behavior without human intervention in the decision-making process.

9.1.3 Experimental results

To verify the effectiveness of the proposed method in this chapter, human–machine collaborative disassembly simulation experiments were conducted on three types of battery packs while ensuring human–machine safety. For A-model aging EVB (EVB-A) pack, a collaboration disassembly strategy was generated based on reinforcement learning method. For B-model aging EVB (EVB-B) pack, the similarity between its disassembly task and that of EVB-A was evaluated, and then a human–machine collaborative task planning was carried out based on transfer reinforcement mixed-driven strategy. Finally, human–machine collaborative disassembly operation was conducted on EVB-C pack in a real operational space.

9.1.3.1 Collaborative strategy generation for EVB-A

The main components of EVB-A pack are shown in Fig. 9.13. Firstly, the 3D model file of the EVB-A is parsed to extract the list of parts contained in the product model, as well as information on the fit and position interference between parts, and to establish constraint relationships between physical objects and geometric features. Secondly, the operation process and difficulty attributes of each component are defined, and the disassembly knowledge map is constructed by integrating part constraint features. Finally, the task status is inputted into the reinforcement learning network, and a collaborative work sequence between human and robot is outputted based on iterative learning of the dual intelligent agents. Part of the sequence of the collaborative disassembly strategy for EVB-A pack is shown in Fig. 9.14.

The overall effect of the collaborative task on the performer is shown in Fig. 9.15, and the complexity value is magnified by a factor of 10 to observe the differences more clearly. Overall, through the designed algorithm, the robot is more inclined to perform relatively complex tasks while minimizing the overall task time. At the same time, in the case of collaboration, the total operation time and complexity of the human are generally smaller than that of the robot.

9.1.3.2 Transfer reinforcement of collaborative strategy for EVB-B

As shown in Fig. 9.16, the EVB-A pack is mainly composed of 13 types of components, while the EVB-B pack is composed of 14 types of elements. Although the two models of battery packs have a high similarity in appearance and structural composition, there are certain differences in terms

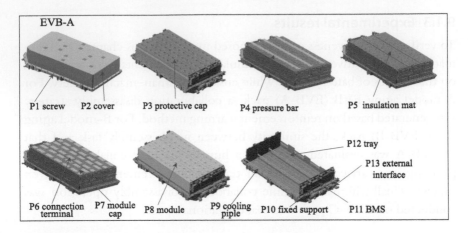

Figure 9.13 The main components of the EVB-A.

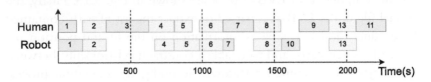

Figure 9.14 Collaborative dismantling strategy for EVB-A.

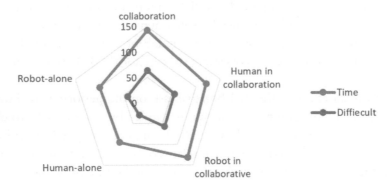

Figure 9.15 Collaboration preference assessment based on time and complexity.

of part size, quantity, and operation sequence. Therefore, in this case, the operational attribute information of the EVB-B pack is first established, and then the constraint information in its model is integrated to construct the disassembly knowledge graph. Subsequently, by discerning the

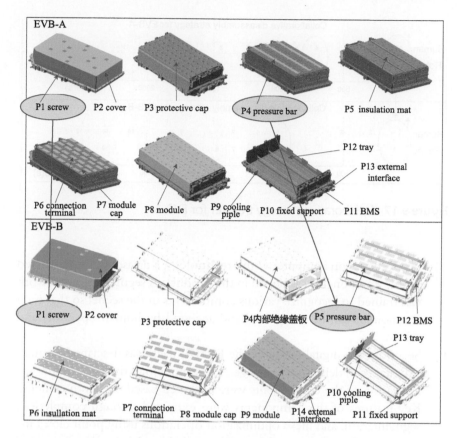

Figure 9.16 Comparison between the composition of EVB-A and EVB-B.

task similarity between the two models of battery packs, the generation of human–robot coadaptive collaboration sequence is conducted based on transfer reinforcement strategy.

The sequence of disassembly planning for the EVB-B pack based on transfer reinforcement method is shown in Fig. 9.17. It is found that the two models of battery packs have identical functions and process characteristics in other parts of the structure except for structural component 4, and their collaborative strategy types can be directly reused.

9.1.3.3 Collaborative disassembly for EVB-C

A disassembly experiment was conducted on EVB-C pack in a real environment, and the experimental process was carried out from three key aspects of the disassembly process:

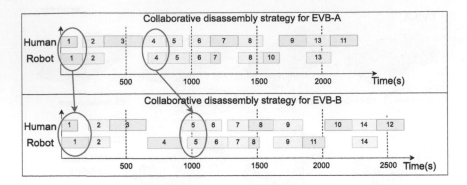

Figure 9.17 Collaborative strategy transfer for EVB-A.

a) Firstly, based on computer vision technology, a dataset was constructed for the components contained in the product. A segmentation network was trained to recognize various components in the scene, so that each component can be recognized and accessed by human operators and robots.

b) Secondly, the digital model of the EVB-B pack was described through a digital twin definition language. The point cloud data and feature information of the components were obtained based on depth cameras. Key features of the components were extracted through completion, segmentation, and mesh optimization operations using point cloud information for reverse modeling. The digital model was then imported into the twin workspace and registered based on virtual and real pose transformations, with the digital programming control module of the virtual model serving as the kinematic driver interface of the twin model.

c) During the execution stage of the collaborative disassembly operation, an operation sequence based on time constraints for human–robot collaboration was outputted through algorithms. The timer was opened when the operation type was outputted, indicating that the task had started, and the status of the robot was updated to occupied.

The resulting operation sequence for the collaborative disassembly of the EVB-C pack is shown in Fig. 9.18.

Without considering the internal structure of the BMS system and module structure, the battery pack used in this experiment consists of a total of 810 components. Due to the lengthy duration of the complete disassembly process and the inability to guarantee absolute recognition ac-

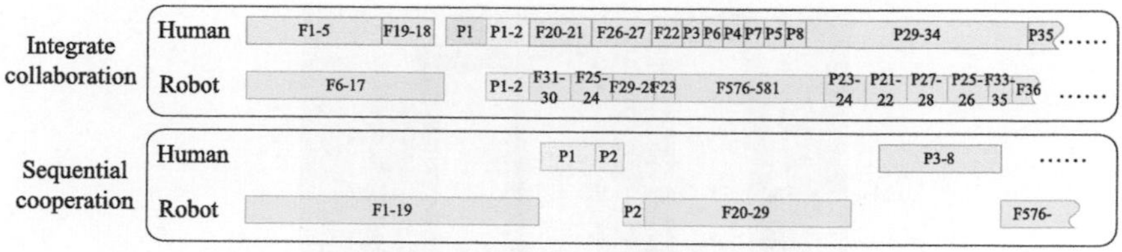

Figure 9.18 Integrated collaborative and sequential cooperation.

curacy during the robot's participation in real collaborative tasks, various interferences and collisions are prone to occur. Therefore, based on experimental constraints and requirements for decision accuracy and operational safety, the robot only performs partial behavioral actions to achieve an illustrative effect. Moreover, due to the complexity of the proposed task planner, the operational details in the task process have been simplified, and a series of operational behaviors are limited to five types: turning loose, lifting, prying, picking up, and pulling out. The completion time for each task was obtained through preliminary testing. The turn-taking time occurring during human–robot interaction and the transition time between different tasks have been included.

To further enhance experimental efficiency and avoid excessive repetition, a task decomposition and reassignment approach was employed to select representative tasks for disassembly validation. The simplified disassembly of the EVB pack consists of 17 subtasks. Tasks 5 and 10 were identified as potentially unsafe for human operators due to the risk of accidental short circuits during manual lifting of the cover or removal of high-voltage cable screws, which could lead to personal injury. Additionally, as shown in Fig. 9.19, task pairs (Tasks 7–9) and (Tasks 11–15) do not meet the minimum safety distance requirements. These task pairs were formulated as subsets of subtasks that can only be performed by the same operator.

The Gantt chart depicting the final disassembly sequence for the 17 components is shown in Fig. 9.20. After task initialization, the operator and robot perform Task 1 together, after which they proceed with individual operations for Tasks 2 and 5, respectively. As Tasks 2 and 3 employ the same tool, the operator can execute this task within a continuous task interval without the need for tool retrieval and replacement. Regarding Task 17, although it is only constrained by Task 13, based on the minimum safety distance criterion, human involvement is not permitted during robot execution.

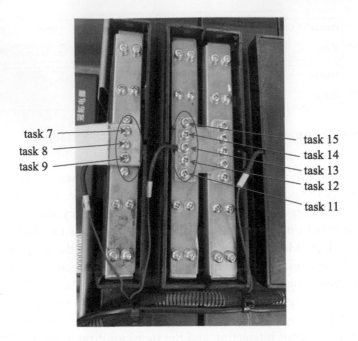

Figure 9.19 Task pair segmentation based on minimum safe distance.

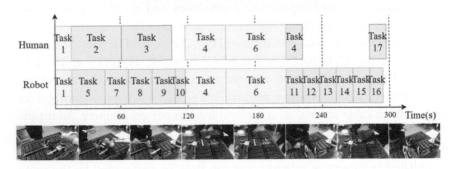

Figure 9.20 Gantt chart of EVB-C disassembly sequence.

9.1.4 Analysis and discussion

Four sets of disassembly control experiments were conducted for the EVB-C, manual/human-only disassembly, robot-only disassembly, human–robot sequential cooperation disassembly, and disassembly based on the human–robot integrated mode. In the stand-alone disassembly mode, all disassembly actions are required to be completed by a single operator using relevant tools. The human–robot sequential cooperation mode prohibits humans

Table 9.1 Results comparison for different modes of operation.

Type	Task completion	Total operating time	Human workload intensity	Human participation rate
Human-only	100%	6767 s	Heavy	100%
Robot-only	87%	4955 s		
Sequential cooperation	100%	5516 s	Low	66%
Integrate collaboration	100%	3835 s	Moderate	43%

and robots appearing in the workspace at the same time, with one party only allowed to intervene in the work after the other completes the operation and exits the workspace. In the case of disassembly using the human–robot integrated collaboration mode, both humans and robots are allowed to simultaneously perform disassembly operations on the battery pack.

The experimental results of each operating mode are compared in Table 9.1. For the human-only disassembly task, it has the longest total operation time and the highest corresponding workload, making it the least efficient task mode. In general, prolonged periods of work can result in fatigue accumulation for human operators, which can affect the execution of subsequent batches of work. Without considering the time consumed in robot preprogramming, the time cost of the robot-only operation mode is the lowest, as there is no human involvement and the robot can operate at a higher speed.

In the aspect of human–robot collaborative operation, the sequential cooperation mode between humans and robots significantly reduces human involvement while compensating for the limitations of robots being unable to independently complete all tasks. The sequential cooperation mode ensures human safety to the greatest extent. As for the integrate collaboration mode, some collaborative tasks can be partially delegated to robots, further reducing human involvement. At the same time, the mode that allows humans and robots to work in the same workspace simultaneously greatly improves task execution efficiency, approaching 35%.

9.2 Collaborative drilling of aircraft skin

The aerospace industry stands as a leadership in manufacturing. One critical phase in the aircraft manufacturing process is the drilling process of aircraft skin during aircraft assembly (Jayaweera and Webb, 2007). The quality of the drilling process impacts the overall lifespan and the durability of the aircraft.

A typical aerospace product is composed of thousands of mechanical components. These components are characterized by labor-intensive processes, a substantial count, intricate system structures, complex external configurations, and diverse material options. Hence, it is difficult to balance manufacturing costs, production timelines, and overall quality during the drilling process.

In traditional production systems, the drilling process of aircraft skin relies on manual operations. This procedure involves several steps, including hole-marking, initial countersinking or hole formation, final countersinking, and subsequent cleaning. In modern factories, traditional manual drilling techniques have been replaced by automated drilling systems (Devlieg, 2011; Zou and Liu, 2011). These automated drilling systems can handle clamping, drilling, counterboring, adhesive application, riveting, and milling in a single setup, reducing processing time and enhancing repeatability.

However, the utilization of such automatic drilling device often demands a long preparatory time for equipment status verification during operation. Real-time monitoring of the current drilling status and quality data is not accessible, leading to issues such as inconsistent hole quality, aperture deviations, and inadequate edge distances. Making adjustments to process parameters of the drilling equipment is not feasible due to the time delays. The integration of cutting-edge technologies, including AR, big data analysis, and AI, to enable immersive automated hole-drilling through Proactive HRC, emerges as a vital technology for ensuring online drilling and drilling quality optimization. The human–robot collaborative drilling of aircraft skin deserves extensive exploration to improve hole-drilling quality performance and ensure aircraft durability.

9.2.1 Workflow of Proactive HRC drilling

Proactive human–robot collaborative drilling is characterized by an immersive interactive operation between drilling robots and human operators throughout the drilling task of aircraft skin. This collaborative system includes the sharing of information concerning drilling conditions and the optimization of drilling process parameters. The overall collaborative drilling process is structured into four sequential modules: 1) real-time perception of the drilling process through the utilization of multiple sensors;

2) online data-driven prediction of drilling results; 3) optimization of the drilling process parameters; and 4) interaction and immersive visualization to facilitate traceable information sharing for the whole process. These modules ensure a seamless collaborative drilling experience, as shown in Fig. 9.21.

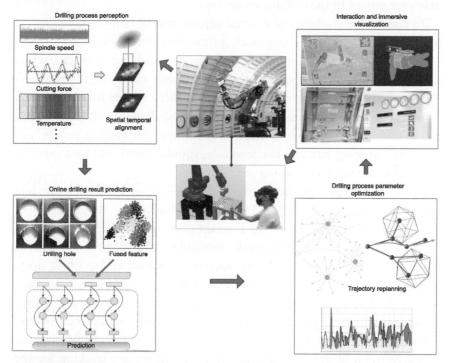

Figure 9.21 The workflow of proactive human–robot collaborative drilling for aviation assembly.

For hole-making quality detection, a real-time perception system can utilize multi-sensor fusion approaches through hole-making processes. It aims for real-time monitoring system of hole-making procedures by considerating hole-making materials, hole quality, and industry standards. To achieve it, we introduce a spatial-temporal approach for hole-making processes using multi-sensors, which integrate visual, vibrational, auditory, and power-based signals. It establishes a method for the spatio-temporal alignment of multimodal perception information. This enables automatically gather multi-dimensional spatio-temporal data of hole-making processes in real time.

Then, an online prediction method can be used to predict drilling results of aircraft skin. This method relies on the utilization of big data and expert knowledge. A multimodal estimation algorithm is leveraged to fuse sensory and visual signals to deduce the quality of the drilling process. The online assessment of hole-making quality can provide real-time alerts to the parameter related to the drilling procedure.

To address the challenge of delayed adjustments in hole-making process parameters, an optimization approach is introduced based on incorporation of multiple inputs and outputs. This method can identify critical factors that impact hole-making quality by simultaneously clarifying the relationship between process status signals and hole-making quality indicators, such as burrs, aperture variations, and edge distances. Based on drilling quality thresholds, the optimization strategy is introduced for drilling process parameters, involving variables like feed rate, spindle speed, and feed speed as outputs, enhancing hole-making precision.

Lastly, an interaction and immersive visualization platform with humans in the loop is developed. This system involves the development of data visualization technology that leverages the Internet of Things (IoT) and AR technologies to automatically recognize drilling processes, provide real-time assessments of equipment operational status, and evaluate hole-drilling quality of aircraft skin in different batches. An interface is designed in wearable AR devices, enabling an immersive and all-encompassing process monitoring and inspection experience. This method also incorporates a process tracing technology for the entire drilling process, facilitating the precise retrieval of hole-making quality data and process tracking for the generation of inspection quality alerts.

9.2.2 Real-time perception of the drilling process

The perception of multiple quality parameters in the robot drilling process typically utilizes a range of sensors, including visual sensors, vibration sensors, sound sensors, force sensors, etc. For example, Yuan et al. (2014) employed laser distance sensors placed at four noncoplanar points evenly distributed around the drilling point. They utilized the center of the circumscribed sphere and the line connecting it to the drilling point to represent the normal vector of the drilling point surface. Alternatively, Yu et al. (2019) utilized machine vision to evaluate the precision of countersink holes and developed a process model for countersink imaging.

Propelled by AI, multimodal information fusion, and data-sharing platforms, real-time monitoring technology in automated robot drilling processes is progressing toward intelligent, comprehensive perception, and cloud-based monitoring mode. Frommknecht et al. (2017) designed a hole-making robot based on multisensor measurement for aerospace product. This robot employs a KUKA KR210 industrial robot equipped with an end effector for visual measurement, normal detection, and precision compensation. Then, during rigid contact in human–robot interactive drilling, Guler et al. (2022) introduced real-time adaptive adjustments of robot admittance controller parameters using a model of artificial neural networks to reduce oscillations during drilling.

Despite these advancements, problems persist in real-time monitoring during robot drilling processes for aircraft products. For instance, the hole-making process often occurs in complex industrial environments where it may be affected by factors such as vibrations, noise, and other disturbances. These interferences can potentially impact the accuracy and stability of sensors. Efficiently fusing and integrating information from multiple sensors in such environments remains a challenge. Addressing this issue hinges on the development of more efficient and accurate multimodal information fusion algorithms.

In response to this situation, a monitoring system for the robot drilling process is developed, as depicted in Fig. 9.22. This monitoring system offers real-time perception of the conditions of the robot, the tool, and the workpieces of aircraft skin, encompassing visual, vibration, structural, electrical, and Radio Frequency Identification (RFID) signals. To manage the lifecycle of tools, RFID technology is employed to establish a system for tool identification, tracking, and management. RFID tags are used for storing and transmitting data related to tool inventory management, records of tool issuance and return, location tracking, and maintenance information. For monitoring the status of robotic equipment, electrical signal sensor is utilized to collect data on robot coordinates overall task execution status.

To perceive the hole-making process, firstly, a profile scanner is employed to swiftly acquire 3D data from the aircraft skin's surface, facilitating the reconstruction of detailed information about the workpiece's shape. This enables high-precision measurement of hole geometry dimensions and spatial distribution. Then, a visual camera is deployed to capture high-frequency image data during the hole-making process, facilitating real-time assessment of hole characteristics, including edge features, hole

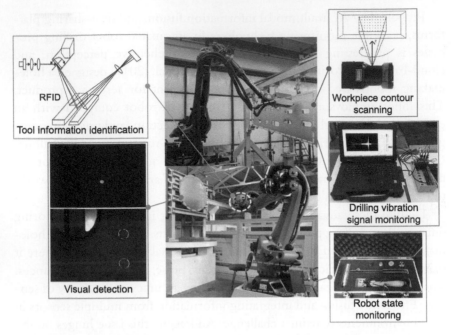

Figure 9.22 The perception system setting of the robot drilling process with multiple sensors.

depth, aperture variations, and burrs. Moreover, fiber optic vibration sensors are utilized to record vibration signals from the aircraft skin during the hole-making process. These signals are analyzed to detect variations in drilling quality and identify abnormal hole-making processes. Lastly, force and temperature sensors are employed to capture data concerning forces and temperature changes during the hole-making process.

Drawing upon the collected process information, which includes spindle speed, cutting force, power, temperature, and vibration, CNN is initially used to extract features from different modal data. Subsequently, these extracted features undergo temporal alignment to synchronize data from various modes within a consistent timeline of robot drilling operations. Then, spatial alignment of the extracted features is performed to associate multimodal data with the different stages of robot, tool, and hole-making formation within the same timeline. Through the use of aligned temporal and spatial data, anomalies in the hole-making process can be detected. In this way, efficient and accurate real-time perception of the hole-making process is achieved, as shown in Fig. 9.23.

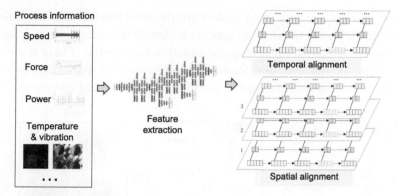

Figure 9.23 The method for real-time perception of the drilling process by multi-sensor fusion.

9.2.3 Online prediction of drilling result

For now, the drilling quality evaluation relies on post-production or offline assessments. Existing methods for evaluating hole-making quality primarily concentrate on single-modal defect detection and classification after the drilling process has concluded. For instance, Cui et al. (2022) devised an algorithm for hole location extraction and a quality assessment method after the drilling operation, which utilized contour extraction, robust circle fitting, and adaptive threshold segmentation. This method also considered the low contrast, morphological variances, and random distribution characteristics of drilling layers in large composite aircraft components. Tang et al. (2022) introduced a framework for automating hole detection in composite flat components, incorporating laser scan data and synthesized scan data to expedite hole position retrieval through circle fitting. Additionally, carbon fiber composites, a fundamental material in aviation, exhibit distinct defects like delamination and uncut fibers during machining processes. In response to this, Hernandez et al. (2020) used visual servoing technology to inspect holes in flat carbon fiber composite panels in aerospace applications. They automated the camera's movement to capture high-resolution images of each hole, thus facilitating the determination of geometric errors and delamination factors. Hrechuk (2023) took into account various material types, tool usage, and cutting conditions, recognizing that the forms and sizes of defects can be unique. An approach for detecting hole defects in polymeric materials is presented, which could be used for high-speed steel drill bits with different cutting data.

Nevertheless, most of these studies are oriented toward assessing hole-making quality after the drilling process is completed, often when defects have already occurred, rendering the situation irreversible. There is a notable dearth of research that focuses on preventive or predictive online assessments of hole-making quality during the hole-making process itself.

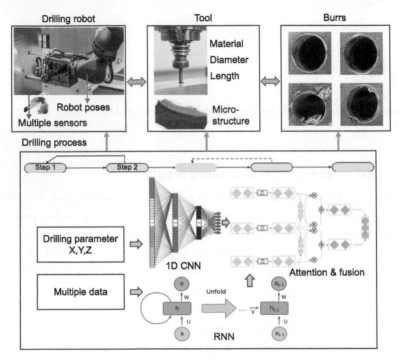

Figure 9.24 A data-driven approach for online analysis of robot drilling results.

Motivated by this research gap, an online predictive approach for robot drilling results of aircraft skin is proposed, as depicted in Fig. 9.24. This approach establishes a connection between the robot's pose, monitored multi-modal data, tool parameters (e.g., wear and microstructure), and hole-making quality. Firstly, 1D CNN and RNN networks are employed to extract features from multi-modal data. The 1D CNN is well-suited for analyzing time series data from sensors and is effective at examining signals with fixed-length periodicity, thus capturing spatial oscillations in drilling and variations in machining forces stemming from drilling parameters. On the other hand, the RNN with temporal memory can capture key feature information from sequences of arbitrary length. It represents semantic features during the process of fusing multiple signals. Through the integration of attention mechanisms and feature fusion, it effectively predict hole-making issues such as burrs, aperture fluctuations, and short-edge distances based on signals of the drilling status, enabling real-time hole-making quality warnings.

9.2.4 Optimization of drilling process parameters

The robot operation parameter optimization is the key to ensuring the machining quality, especially in drilling, milling, grinding and other manufacturing tasks. For example, Liao et al. (2021) developed an approach for optimizing robot posture and workpiece setup in robotic-assisted milling tasks. An index of the robot's rotational deformation was proposed to evaluate the robot stiffness by considering robot redundancy. Lin et al. (2022) proposed a task-dependent pose optimization method which considers the milling process and contour error-based machining performance index. The tooltip deformation errors along the machining path are assessed based on the model of robot stiffness and simulation-based cutting forces, then the contour errors were tested according to the deformation errors and the working path profile. Chen and Ding (2023) presented an effective posture optimization method for robotic flat-end milling. The inclination, tilt angles, and robot redundancy at every working point on a milling path are optimized simultaneously to improve the path smoothness and machining width.

Nevertheless, most of these works focus on parameter optimization after the manufacturing process is completed, which is not feasible and time-efficient. Thus, developing some methods for real-time or automatic drilling parameter optimization has attracted attentions. The critical issue is how to achieve real-time and optimal parameter improvement to enhance the robotic drilling quality based on drilling quality evaluation factors (Chen et al., 2023).

Driven by this key challenge, one hybrid parameter optimization method is developed, as shown in Fig. 9.25. In this process, four different quality assessment criteria (drilling efficiency, export burr, hole size, and inner scratches) are applied in this approach to reflect the drilling level of the drilling task on aircraft skin. Then, we combined Particle Swarm Optimization (PSO) and case-based reasoning optimization to select the optimal/near-optimal drilling parameters. This approach builds feedback between drilling optimal/near-optimal factors and hole-making quality testing. According to this feedback, the drilling parameters, such as robotic

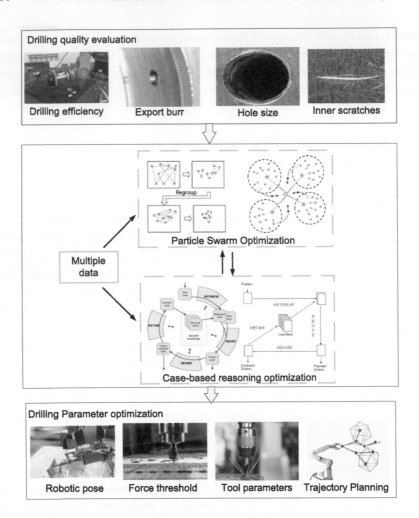

Figure 9.25 The workflow of optimization of drilling process parameters.

pose, drilling force, trajectory planning, and tool parameters, can be adjusted automatically and rapidly.

9.2.5 Interaction and immersive visualization

In industrial spot-checking scenarios, AR technology provides opportunities for intelligent interaction, expert remote assistance, and seamless access to knowledge repositories. For example, Marino et al. (2021) utilized AR tools with a virtual-physical fusion feature to superimpose 3D models onto physical assembly objects, enabling inspection personnel to easily detect design differences in the final assembled products. The use of this AR tool not only reduced cognitive demands but also garnered high satisfaction among factory workers and engineers participating in user studies. Runji and Lin (2020) introduced an AR-based human–robot collaborative drilling scenario, combining automated optical inspection with wireless assistance and SLAM for real-time tracking of assembly progress. Additionally, Liu et al. (2019) implemented AR technology for equipment operation monitoring and maintenance, simplifying the complexities associated with workshop equipment management and reducing the expertise required for maintenance personnel. By creating an AR monitoring and maintenance system that merges virtual and real environments, it presented information to humans, guiding their maintenance tasks and enabling the visual management of production site equipment.

Hence, AR technology provides natural and immersive monitoring and information tracing in human–robot collaborative drilling scenarios. For example, it can assist individuals in conducting real-time spot inspections for hole-making quality and operational equipment. Furthermore, AR devices can function as display terminals for the IoT, integrating human operators into cyber-physical systems. By incorporating big data technology, AR can support proactive human–robot co-working and intelligent teamwork, facilitating the real-time traceability of hole-making process information.

In this context, we introduce an interactive drilling and traceable information visualization system in an AR environment to enhance human context awareness of the whole process of aircraft skin drilling, as depicted in Fig. 9.26. In a proactive human–robot collaborative drilling setting, virtual CAD models can integrate with physical workstations of aircraft skin through AR tracking and registration technologies. This system combines virtual and real environments, facilitating coordinated interactions between humans and drilling robots, thereby enhancing their proactive collaborative capabilities in complex hole-making processes. On the one hand, it involves human–robot interactions during the drilling process. A Transformer model can be used to recognize human intention from speech, gestures, and eye movements, for high-precision, multimodal human–robot communication. A cloud-edge collaborative system is deployed to alleviate computational burdens associated with the speech, gesture, and eye movement recognition. On the other hand, the system includes the development of traceability and retrieval database for the entire drilling process for aircraft skin in changing batches, storing knowledge representations of spot-check

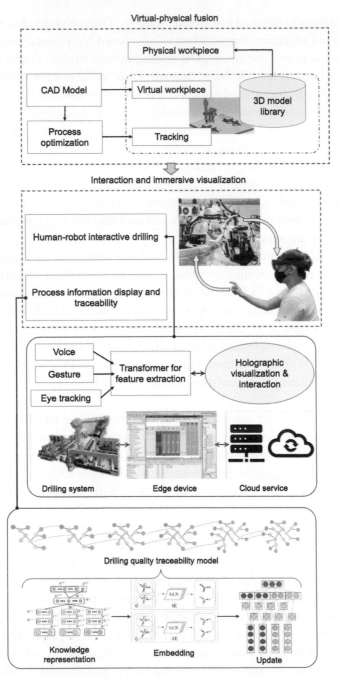

Figure 9.26 Interactive drilling process and traceable information visualization in AR environment.

quality information in a knowledge graph. Through various node embedding calculations and link prediction techniques, it updates the knowledge graph's representations for different process flows and presents them to users in an AR environment.

References

Chen, H., Huo, S., Muddassir, M., Lee, H.Y., Duan, A., Zheng, P., Pan, H., Navarro-Alarcon, D., 2023. PSO-based optimal coverage path planning for surface defect inspection of 3C components with a robotic line scanner. arXiv:2307.04431.

Chen, Y., Ding, Y., 2023. Posture optimization in robotic flat-end milling based on sequential quadratic programming. Journal of Manufacturing Science and Engineering 145, 061001.

Cui, J., Liu, W., Zhang, Y., Han, L., Yin, P., Li, Y., Zhou, M., Wang, P., 2022. A visual inspection method for delamination extraction and quantification of carbon fiber reinforced plastic (CFRP). Measurement 196, 111252.

Devlieg, R., 2011. High-accuracy robotic drilling/milling of 737 inboard flaps. SAE International Journal of Aerospace 4, 1373–1379.

Frommknecht, A., Kuehnle, J., Effenberger, I., Pidan, S., 2017. Multi-sensor measurement system for robotic drilling. Robotics and Computer-Integrated Manufacturing 47, 4–10.

Guler, B., Niaz, P.P., Madani, A., Aydin, Y., Basdogan, C., 2022. An adaptive admittance controller for collaborative drilling with a robot based on subtask classification via deep learning. Mechatronics 86, 102851.

Hernandez, A., Maghami, A., Khoshdarregi, M., 2020. A machine vision framework for autonomous inspection of drilled holes in CFRP panels. In: 2020 6th International Conference on Control, Automation and Robotics (ICCAR). IEEE, pp. 669–675.

Hrechuk, A., 2023. Recognition of drilling-induced defects in fiber reinforced polymers using machine learning. Procedia CIRP 117, 384–389.

Jayaweera, N., Webb, P., 2007. Adaptive robotic assembly of compliant aero-structure components. Robotics and Computer-Integrated Manufacturing 23, 180–194.

Liao, Z.Y., Wang, Q.H., Xie, H.L., Li, J.R., Zhou, X.F., Pan, T.H., 2021. Optimization of robot posture and workpiece setup in robotic milling with stiffness threshold. IEEE/ASME Transactions on Mechatronics 27, 582–593.

Lin, J., Ye, C., Yang, J., Zhao, H., Ding, H., Luo, M., 2022. Contour error-based optimization of the end-effector pose of a 6 degree-of-freedom serial robot in milling operation. Robotics and Computer-Integrated Manufacturing 73, 102257.

Liu, L., Jiang, C., Gao, Z., Wang, Y., 2019. Research on real-time monitoring technology of equipment based on augmented reality. In: Advanced Manufacturing and Automation VIII. Springer, pp. 141–150.

Marino, E., Barbieri, L., Colacino, B., Fleri, A.K., Bruno, F., 2021. An augmented reality inspection tool to support workers in Industry 4.0 environments. Computers in Industry 127, 103412.

Runji, J.M., Lin, C.Y., 2020. Markerless cooperative augmented reality-based smart manufacturing double-check system: Case of safe PCBA inspection following automatic optical inspection. Robotics and Computer-Integrated Manufacturing 64, 101957.

Tang, H., Zhou, L., Liu, Y., Wang, J., 2022. Tiny hole inspection of aircraft engine nacelle in 3D point cloud via robust statistical fitting. Measurement 196, 111250.

Yu, X., He, W., Li, Y., Xue, C., Li, J., Zou, J., Yang, C., 2019. Bayesian estimation of human impedance and motion intention for human–robot collaboration. IEEE Transactions on Cybernetics.

Yuan, P., Wang, Q., Shi, Z., Wang, T., Wang, C., Chen, D., Shen, L., 2014. A micro-adjusting attitude mechanism for autonomous drilling robot end-effector. Science China. Information Sciences 12, 1–12.

Zou, C., Liu, J., 2011. An off-line programming system for flexible drilling of aircraft wing structures. Assembly Automation 31, 161–168.

CHAPTER 10

Conclusions and future perspectives

Based on the elaborative exposition of Proactive HRC in the previous chapters, the concluding chapter summarizes the key discoveries and contributions made in this book. The literature analysis, connotation, methodology, and demonstration of Proactive HRC play a crucial role in today's ever-increasing transition of HCSM environment. Meanwhile, multiple pressing challenges and future research directions of Proactive HRC are highlighted here to welcome more open discussions and in-depth research and development. These encompass a wide array of dimensions, such as the evolution of its intelligent capabilities, the human-in-the-loop control, and the development of effective evaluation criteria.

10.1 Conclusion

The manufacturing industry is actively pursuing a paradigm shift toward flexible automation and enhanced collaboration between human workers and machines. This drive is motivated by the pressing challenges of an aging workforce, labor shortages, and a growing demand for individually tailored products at scale. In response to these imperatives, the concept of HCSM is gaining prominence. HCSM represents a fusion of cutting-edge technologies, including artificial intelligence, robotics, and the industrial Internet of Things, with human-centric design principles. The aim is to forge a production environment that is both collaborative and adaptive. In essence, HCSM underscores the pivotal role of humans in the factories of the future. The overarching objective is to amplify productivity, enhance safety, and elevate the well-being of operators while concurrently elevating the efficiency, flexibility, and quality of production processes.

In the context of HCSM, HRC emerges as a promising and pivotal solution to boost manufacturing performance (Chapter 1). HRC hinges on harnessing the distinctive strengths of both human workers, characterized by cognitive flexibility and adaptability, and robots, known for their exceptional precision, strength, and repeatability. The ultimate goal of HRC is to create complementary of these attributes to drive produc-

Proactive Human–Robot Collaboration Toward Human-Centric
Smart Manufacturing
https://doi.org/10.1016/B978-0-44-313943-7.00017-X

tivity and enhance worker satisfaction. Extensive research endeavors have been undertaken to develop a diverse array of HRC systems and technologies. These innovations facilitate seamless and efficient collaboration between humans and robots. Key components include sensing and perception systems, decision-making methodologies, user-friendly interfaces, and advanced planning and control algorithms. These HRC systems can be flexibly applied across a spectrum of manufacturing tasks, encompassing functions such as machining, welding, and assembly. They afford human operators the opportunity to acquire new skills and engage in tasks requiring problem-solving and creative prowess. Simultaneously, robots shoulder repetitive, hazardous, or physically demanding tasks, contributing to an uptick in productivity and the overall well-being of workers.

Nevertheless, the current landscape of HRC applications confronts a series of challenges that impede the realization of optimal synergy between human and robot capabilities. These challenges manifest in the following dimensions:

- (*Limited Adaptability*) Present-day HRC systems are often tailored for specific environments, lacking the essential flexibility needed to seamlessly adapt to evolving work conditions and dynamic task requirements.
- (*Weak Context Awareness*) The contextual awareness embedded within HRC systems tends to remain at a nonsemantic perception level. This limitation hinders their capacity to comprehensively understand the nuances of the work environment and the intricate interplay between human, robot, and task.
- (*Ineffective Decision-Making*) Many contemporary HRC systems heavily rely on preprogrammed instructions, curtailing their ability to make real-time decisions and proactively adapt to changing work conditions and requirements.
- (*Poor Trust*) The bi-directional understanding between humans and robots is lacking, as humans may not fully trust the capabilities of robots, and robots may not fully comprehend human intentions.

In response to the compelling challenges outlined above, this book embarks on an exploration of a visionary paradigm, referred to as **Proactive HRC**. This paradigm is aligned with the ongoing transformation of HCSM, presenting a forward-looking framework for addressing the intricate dynamics of modern manufacturing. Traversing along the trajectory of 5C intelligence involvement, Chapter 2 undertakes a comprehensive examination of human–robot relationships that transpire within the manufacturing domain. This journey takes us through the evolutionary phases of coexistence, interaction, cooperation, and ultimately, collaboration. Each phase is dissected, with a focus on delineating their respective attributes, characteristics, and inherent limitations. The overarching aim is to provide a well-rounded perspective on the dynamic requirements within each stage of human–robot teamwork. Overarching principle of human-centric design and operational requirements, the concept of Proactive HRC is introduced. It stands for the drive of a heightened level of flexible automation that seamlessly integrates into the manufacturing environment. At its core, Proactive HRC embodies bi-directional proactivity, where humans and robots engage collaboratively drive manufacturing activities with a focus on responsiveness, adaptability, and mutual enhancement of productivity. This vision reimagines the role of humans and robots as proactive teammates, participating in a flexible and efficient manufacturing task.

Chapter 3 delves into an in-depth exploration of the architectural underpinnings, the conceptual framework, and the pivotal enabling technologies that together constitute the Proactive HRC paradigm. Proactive HRC embodies a aspiration: to cultivate intellectual capabilities that are similar to human cognition while simultaneously fostering proactivity in collaborative manufacturing tasks within both humans and robots. Central to the implementation of Proactive HRC systems are four indispensable modules, each of which contributes its unique dimension to the proactive collaboration:

- (*Mutual-Cognition and Empathy*) This module means the mutual-cognitive understanding ability and empathic robot skills in Proactive HRC. It enables participants to grasp each teammate's needs and align to time-changing task requirements.
- (*Predictable Spatio-Temporal Collaboration*) The pursuit of predictable spatio-temporal collaboration allows for long-term fluent task fulfillment and foreseeable interactive collaborations in the execution loop.
- (*Self-Organizing Multi-Agent Teamwork*) In self-organizing multi-agent teamwork, multiple humans, in unique roles, and a diverse array of robotic agents, with distinct workloads, converge within complex manufacturing tasks. This multi-agent system allocates manufacturing resources and plans task arrangement in an explainable manner.
- (*Intelligent Robot Control and Human Assistant System*) The intelligent robot control and a human assistant system ensure that robots operate with generalization and transferability, while human operators are equipped with assistance systems that amplify their abilities and offer essential support in intricate manufacturing tasks.

For the mutual-cognitive intelligence and empathetic skills, Chapter 4 explores an MR and visual reasoning method in Proactive HRC. It can achieve scene interpretation and the strategic assignment of task arrangements and consider the bi-directional operational goals of humans and robots in manufacturing tasks. Within this framework, the robot motion adheres to ergonomic principles when interacting with humans. Simultaneously, the human participants can obtain intuitive information support from the MR environment. This enhances human perceptual abilities in the collaboration for a global understanding of manufacturing tasks. This proposed system propels HRC systems beyond the perceptual confines. Proactive HRC is a cognition system, where human and robotic agents can understand knowledge of the environment, tasks, and collaborative dynamics. This cognitive transformation empowers the human participants to exhibit high-level intuitive behaviors, while concurrently ensuring the proactive role of robots in co-working.

Chapter 5 introduces a multimodal transfer learning algorithm to anticipate human operation intentions during HRC in advance for the predictable spatio-temporal collaboration. Proactive robot motion thereby is achieved with awareness of human future intentions, which ensures a seamless and predictable collaborative environment. A multimodal fusion network is the key to accurately predicting human next intentions in the immediate future. It excels even when fine-grained operations, such as wedging pins and screwing bolts, present patterns that might otherwise confound the predictive process. Besides, the network shows adaptability in different datasets when predicting ongoing human intentions with precision. For robot motion planning, a decision-tree mechanism is used to link human intentions and robot actions.

Chapter 6 introduces the self-organizing task planning in Proactive HRC. This approach tackles limitations associated with predefined task arrangements, offering a dynamic collaborative process in HRC. The foundational mechanism is temporal subgraph reasoning – that learns knowledge representation of task planning between multiple human operators and robotic agents. This mechanism fuses knowledge of prior experiences, the progression of task stages, and the dynamic operating environment to forecast the next operational steps for each agent, paving the way for an agile and adaptable collaborative process. Simultaneously, the temporal subgraph structure provides a transparent and interpretable graphical framework. Within this framework, hierarchical and sequential task arrangements take shape to facilitate the seamless collaboration between human operators and robots. The graphical representation allows human participants to gain a comprehensive and global understanding of the HRC task, like which task they are situated. For the robotic agents, the temporal subgraph provides motion commands. By parsing the graph triples encapsulating robot–action–component connections, the robotic entities can translate these commands into automatic motions.

Chapter 7 provides roadmap in the development of a Proactive HRC system based on four fundamental modules. These modules encompass scene perception, knowledge representation learning, decision-making, and collaborative control. Within each module, related technologies are discussed and presented in tables. These tables offer a structured overview, providing readers with insights into the technologies that underpin Proactive HRC. Moreover, the chapter delves into prospective solutions with detailed algorithms that correspond to each of the four foundational modules. Noteworthy examples include the concept of a human Digital Twin, the implementation of 3D scene perception, the utilization of knowledge graph-based adaptive task planning, and the infusion of collaborative intelligence into system control. These solutions empower Proactive HRC with a heightened degree of intelligent capability.

Chapter 8 provides typical operational modes of Proactive HRC within industrial contexts. These modes encompass continuously hierarchical and sequential operations, co-manipulation, and physical interaction, all of which are recurrent scenarios in manufacturing task. To facilitate collaborative hierarchical and sequential tasks, the application of AND/OR graph methods is introduced to generate a role allocation strategy representation. In pursuit of optimal graph search, the AO* search algorithm is employed, inspired by the A* algorithm. For scenarios involving comanipulation and physical interaction between humans and robots, the MOCA-MAN interface is developed. This interface comprises an Arduino Nano microcontroller, a force-torque sensor, and an end-effector. The MOCA-MAN interface allows for efficient manipulation, where humans and a mobile manipulator coordinate each other's capabilities effectively. For example, humans assume the roles of both collaborators and educators for mobile manipulators, during the teaching process with their physical interactions.

With these exceptional capabilities, Proactive HRC unfolds its technical and operational prowess in smart manufacturing, as depicted in Chapter 9. Two illustrative applications of Proactive HRC are introduced in this chapter. The first application is in the high-tech renewable energy sector, specifically focusing on the disassembly of aging electric vehicle batteries. In this task, human operators can execute ultraflexible operations, responding to dynamic variables, while robotic counterparts shoulder the mantle for high-risk and high-payload operations, thereby mitigating potential hazards. The second application is in the domain of advanced aviation, where human–robot collaborative drilling plays a pivotal role. This application facilitates the sharing of lifespan information between humans and robots throughout the drilling process. This real-time exchange of data fosters a dynamic environment where online drilling result prediction becomes feasible, optimizing the drilling process for enhanced efficiency and quality. These two representative applications of Proactive HRC integrate human and robot capabilities, providing solutions for other production processes of complex products.

In summary, the evolution of Proactive HRC represents a significant stride in the field of intelligent robotic systems. With dynamic perception, cognitive capabilities, and intelligent control, robots transcend their traditional roles and emerge as entities capable of intelligent decision-making and dexterous manipulation. They seamlessly integrate into the collaborative system, proactively coordinating their efforts with their human counterparts.

This book provides comprehensive insights and practical exemplars of Proactive HRC. Its intention is to kindle meaningful dialogs among researchers across the globe, propelling innovative discussions that push the boundaries of HRC. Furthermore, it endeavors to furnish contemporary enterprises and factories with a practical production framework, one that aligns with the evolving market dynamics, enabling them to stay nimble and responsive to changing demands.

Additionally, this volume stands as an invaluable reference for university students pursuing advanced courses, especially in the domain of industrial engineering. It provides a rich reservoir of knowledge and practical insights, equipping students with a robust foundation in Proactive HRC. In essence, this book bridges the gap between theory and practice, making it an indispensable resource for all those vested in the future of Proactive HRC.

10.2 Challenge and future perspectives

The concept of Proactive HRC represents an imminent paradigm shift, promising a future characterized by mutual trust, intellectual synergy, and adaptable collaborative production. The successful realization of Proactive

HRC in real-world scenarios hinges on the in-depth examination of technical, practical, and ethical aspects, as well as the exploration of pertinent theories. Within this section, we shall delve into a couple of pressing challenges related to the cognition, predictability, and self-organizing attributes in Proactive HRC.

1. Online recognition of both short- and long-range human intention

Recognizing human intentions in manufacturing activities is a pivotal prerequisite for cognition in Proactive HRC. When a human operator collaborates with robots, they behavior both short-term and long-term operational goals. Historically, research endeavors have been dedicated to discerning human physical and psychological behaviors within fixed time sequences. However, these conventional approaches falter when it comes to comprehending dynamic human operation intentions that manifest in uncertain time durations.

This scenario leaves us grappling with unresolved social and technical challenges regarding the progressive online identification of human intentions. For instance, should the HRC system consider all types of human intentions throughout the entire task process, and how to clarify a type of intention? This remains a question without a universally accepted societal consensus. It is worth noting that human actions in the current moment and their subsequent steps almost exhibit few discernible changes for different intentions.

From a technical perspective, addressing these challenges necessitates the development of online human intention recognition algorithms capable of decomposing continuous human actions into discrete substage intentions and determining the optimal junctures for recognizing the subsequent human operational goals. This conundrum further extends into the realm of continuous classification and predictive modeling of these successive substage human intentions across varying temporal horizons.

To implement effective human intention recognition within the framework of Proactive HRC, two promising avenues of research emerge. Firstly, there's an opportunity to extract both explicit and implicit human intention patterns by fusing multimodal data, encompassing visual action patterns, voice commands, fluctuations in EMG signals, and the interpretation of mental states derived from brainwave data. Leveraging abundant information across multiple sensory modalities, it becomes feasible to introduce attention and fusion mechanisms into recognition models to determine the transitional time spans demarcating distinct human operational goals and continuously forecast human intentions within varying temporal scopes during a given task.

The second way revolves around the integration of human intention data and the foundational knowledge underpinning HRC tasks through a KG. Establishing a linkage between human behaviors and the domain-specific manufacturing knowledge can offer guidance and cues for the sequential recognition of human intentions as they evolve over time.

2. A convenient and accessible way for human well-being estimation

Presently, the assessment of human well-being has emerged as a focal point, garnering significant attention due to its pivotal role in enhancing human job satisfaction in Proactive HRC. Manufacturing activities involving human operators and robots in close proximity can exact a toll on both physical and mental equilibrium. Past investigations have often relied upon the utilization of EMG and EEG devices to gauge human muscle workload and monitor psychological states, respectively.

However, this conventional approach frequently mandates the donning of bulky and cumbersome sensor equipment, impairing human agility and flexibility, which, in turn, has an adverse impact on human well-being. The ethical dimension of estimating human well-being should consider the reliability and acceptability of models that correlate human emotional fluctuations with physical and mental states. On the technical front, the challenge lies in introducing an approach to estimate human well-being within HRC tasks that is both convenient and unintrusive, allowing human operators to execute their operations with agility.

Two promising techniques for convenient human well-being estimation deserve exploration. Firstly, a DT approach to modeling human physical fatigue can be achieved through the fusion of human biomechanics with real-time visual data including human postures, movements, and point cloud representations. The creation of a digital surrogate of a human worker facilitates the simulation of load-bearing tasks within HRC systems. Furthermore, ergonomic analysis methodologies can be seamlessly integrated into the framework of human DT.

Meanwhile, with regard to the estimation of human mental stress, it is possible to analyze and infer this state by observing a worker's continuous feedback and responses to their surrounding environment. Parameters such as the frequency of gaze directed at nearby robots can serve as indicators for gauging and comprehending the mental stress experienced by human operators in HRC settings.

3. Human–robot–environment parsing and cognition

In the context of Human–Robot–Environment (HRE) interaction, the dynamic interplay among these elements plays a vital role in guiding resource allocation decisions during collaborative processes. While prior research efforts have yielded feasible solutions for parsing HRE scenes, they only targeted specific scenarios. These explorations rely on monitoring robot movements, estimating object positions, or generating navigation maps. Nevertheless, practical hurdles persist in delineating the interpretive relationships between human activities, robot manipulations, and the surrounding environment. The key lies in discerning which forms of combination between these elements carry significance when generating optimal HRE task arrangements. For example, when human handovers coincide with rapid robot-reaching actions, safety is a paramount concern in HRE systems due to the potential for contact hazards.

To address the HRE parsing and cognition, there is a potential solution by fusing visual reasoning techniques and KG methods. In particular, the employment of visual reasoning modules spanning object recognition, semantic comprehension, and knowledge representation can align with the perceptual results derived from HRE tasks, relations, and latent cognitive processes, respectively. Meanwhile, the KG uses a graphical structure to interpret the contextual relationships among these HRE elements. By linking these entities within distinct structural schemas, the KG approach enable comprehensible knowledge for the decision-making processes during HRE interactions.

4. Mutual-cognitive and empathetic teamwork

The Proactive HRC aims promote mutual cognitive and empathetic teamwork, thereby enhancing the overall experience for both humans and robots. Prior research tries to coordinate robot movements and human tasks based on the understanding of task structures and production objectives. Among these systems, human and robotic agents fail to learn and address the collaborative needs of their teammates and lack empathic skills.

For Proactive HRC, firstly, more studies should discuss the establishment of standardized metrics for assessing the effectiveness of mutual cognitive and empathetic teamwork. This might entail the incorporation of ergonomics criteria, evaluations of psychological factors, and productivity measurements to ascertain the holistic performance of HRC systems. Furthermore, as human operators become equipped with information display devices designed to bolster their domain knowledge and enhance their skill sets, particular attention must be devoted to ensuring the temporal and spatial consistency of these visualization tools. This endeavor can optimize human engagement and immersion levels, thereby fostering mutual-cognitive teamwork between humans and robots.

Lastly, the progression of robot learning modules represents a pivotal frontier in enhancing the capabilities of robots. Robots present empathetic skills by aligning with human-desired operations via path re-planning. This path re-planning approach elevates the adaptability and versatility of robots within human-centric concerns.

5. Efficient fusion of on-site scene understanding and prior knowledge of HRC systems

To facilitate fluent decision-making for forthcoming human and robotic operations, the combination of current scene comprehension and the pre-existing knowledge of HRC tasks holds the key. Prior investigations have explored the knowledge representation of common scenarios and the overarching task structures within HRC settings. Nevertheless, a series of technical problems persist, particularly concerning the integration of real-time scene understanding and how it is connected to prior knowledge and the comprehensive HRC procedures.

To ensure HRC tasks with predictability, the fusion of on-site scene comprehension with manufacturing expertise can use KG methodologies. Firstly, a holistic HRC KG can be established, wherein the core production components, namely "Human", "Robot", "Material", "Method", and "Environment" are established as nodes bearing distinct attributes. Subsequently, the elements currently detected within the given scene are dynamically activated within this comprehensive HRC KG. This activation process is then followed by a search algorithm to optimize graph configurations, for guiding the subsequent human and robot operations.

Conversely, the production elements derived from on-site scenarios can be constructed as an SG. This SG is firstly established as an independent entity, subsequently aligning itself with an HRC KG of the whole process. This alignment process is used to adapt and fine-tune the strategies for updating graph connections, thereby guiding suitable task plans for the next HRC stages.

6. Task re-planning and knowledge transfer strategy when facing unexpected or unseen HRC situations

The mechanism for re-planning tasks plays a pivotal role in mitigating the impact of unexpected situations that arise due to human motion uncertainty in Proactive HRC. It can suspend incorrect robot operations while resuming normal procedural sequences. In addition to its adaptability in handling uncertainties, this mechanism enables the transfer of knowledge learned from past collaborative strategies to generalized task planning, specifically for new yet similar HRC task variants.

For now, research efforts directed toward HRC task planning are centered on the scheduling of specific operations with predefined tasks. These conventional task-planning algorithms fail to assure fluent HRC when confronted with the unpredictability of human actions and the introduction of product variations, which are unavoidable factors in the practical world. Consider the scenario where a human operator deviates from the prescribed task precedence constraints or when new product requirements emerge, it is difficult to achieve the generation of re-planning decisions to ensure the seamless completion of HRC tasks.

In light of these challenges, there are several promising solutions for task re-planning and knowledge transfer in Proactive HRC. Firstly, the establishment of an HRC DT system capable of modeling, simulating, and predicting the forthcoming attention, behaviors, and workload of each agent provides a solution. This predictive capability allows for the early identification of potential human errors and interventions to prevent robot runaway motions. Furthermore, even amid unexpected deviations in task progression, RL models can be deployed to identify globally optimal strategies for re-planning subsequent operations. This entails the integration of simulation processes within the DT environment, enhancing the adaptability and resilience of HRC task planning. Lastly, the creation of TL-based KG can transfer HRC skills into a set of general operational rules that can be applied to new, albeit similar, HRC tasks. This approach ensures a seamless and efficient transference of expertise from past experiences to novel situations, thereby enhancing the adaptability and skill acquisition in Proactive HRC.

7. HRC cell design within configurable and ergonomics principles

A fundamental prerequisite for ensuring a comfortable and user-friendly environment in hybrid human–robot teamwork is the implementation of an HRC cell design that is both configurable and comfortable. Previous studies that have delved into this domain have focused on designing work cells targeted for human operators working alongside fixed robot arms. However, these explorations primarily focus on basic ergonomic requisites and task-specific objectives, without consideration of self-configuration and cognitive ergonomics-based design. This shortfall underscores the persistence of social and technical challenges yet to be fully addressed. For instance, the complex dynamics of shared workspaces necessitate the consideration of resource allocation and occupation management. Meanwhile, it is vital to create an HRC environment that caters to the diverse needs of different human operators, particularly in the social dimension. On the technical front, the design of configurable HRC cells must incorporate a comprehensive understanding of task structures, the ever-changing requirements of the environment, and the unique capabilities of each participating agent.

Looking forward to the broader application of HRC cell design, two distinct paths emerge as noteworthy. First and foremost, the development of cognitive ergonomics metrics holds the promise of both the physical well-being of individuals (including aspects like fatigue and overload) and the psychological dimensions of their experience (such as self-esteem and emotional satisfaction). Ergonomics requirements, task resource availability, and the specific roles that humans and robots perform can be seamlessly integrated into a KG. Within this KG, they are mapped as considerations associated with graph nodes, each endowed with a unique set of attributes. The application of KG methodologies extends to the learning of design requirements derived from these nodes. Concurrently, RL-based generative design techniques hold the potential to directly generate HRC cell layouts. This is achieved through the assimilation of insights from extensive historical design experiences, all while adhering to the constraints imposed by cognitive ergonomics considerations.

8. Extendable human and robotic agents for multiple HRC

The incorporation of additional human and robotic agents into HRC systems is essential for augmenting overall productivity, particularly in the context of intricate tasks. Past research has fixated on addressing the computational consumption and latency challenges that arise when accommodating a larger cohort of agents within HRC systems. However, the development of a seamless plug-and-play network connection to facilitate the integration of expandable human and robotic agents, has received relatively limited attention. On both technical and practical fronts, it is critical for realizing the vision of Proactive HRC with a hybrid ensemble of multi-agents. The complexity is underscored by the need to introduce new human–robot groups into existing configurations, thereby optimizing the utilization of their collective capabilities.

An effective solution to this challenge involves the establishment of extendable and multi-agent Proactive HRC systems, organized around cellular human–robot groups. These cellular groups serve as the foundational building blocks, subsequently interconnected through a plug-and-play network approach that spans the hardware, information, and knowledge dimensions. In terms of hardware connectivity, diverse cellular groups, comprising both human and robotic agents, converge within a standardized IIoT environment through a uniform communication interface. This interconnectivity underpins the sharing of resources and insights while fostering a seamless collaborative system.

In the information domain, data transmission within the IIoT environment capitalizes on cutting-edge infrastructure such as 5G and Wi-Fi 6, thereby minimizing communication latency and ensuring real-time coordination. Each human–robot group is represented by distinct KGs, which encapsulate the unique and qualified operations of every agent within the cellular group. These KGs optimize each HRC group capabilities suited for the task at hand.

Furthermore, for knowledge exchange, the methodology of federated learning is suitable to aggregate, converge, and align knowledge across these distributed human–robot groups. This approach ensures that knowledge learned from various groups is fused, facilitating the collective growth of the broader HRC network. Even in the event of introducing a new cellular group of human and robot agents, their seamless assimilation allows them to partake in task operations well-matched to their individual capabilities, thus optimizing the overall performance and productivity of the multi-agent Proactive HRC system.

9. Industrial embodied intelligence in Proactive HRC

Industry 5.0 and human-centric smart manufacturing prioritize the physical and mental well-being, as well as the safety of individuals, by requiring machines to actively align with human needs and desires. This places a significant demand on the intelligence of machines. Robots need to perceive multimodal instructions from human operators, such as verbal commands and gestures, and engage in natural interactions. They should possess a level of cognition to actively collaborate, as well as perform complex and precise movements reliably. In this context, the industry requires more than just machines capable of simple repetitive tasks; it necessitates industrial embodied intelligence with high-level perception, cognition, and action intelligence. The recent emergence of LLM technology, along with the maturation of robotics, has made industrial embodied intelligence feasible. The remarkable generalization ability of LLM enables natural, unscripted dialogs between humans and LLM, while VLM facilitates multimodal interaction between humans and robots. An exemplary task that allows natural multimodal interaction between humans and AI agent is Vision Language Navigation (VLN), which enables cooperative interaction and coordination of AI agent with the objects in the environment through dialog. In industrial settings, industrial embodied intelligence capable of performing VLN tasks can serve as an assistant robot for human operators. For example, when a human operator is engaged in complex assembly or machining tasks at a workstation and realizes a forgotten tool, but cannot temporarily leave their position due to certain reasons, a robotic assistant can be deployed to locate the tool based on the operator's needs fetch it to the workstation. Apart from VLN, LLM-based industrial embodied intelligence can also facilitate tasks such as task planning and allocation. Thus, LLM-based industrial embodied intelligence holds great promise as a future research direction in HRC for human-centric smart manufacturing.

10. LLM-based Proactive HRC in intricate manufacturing tasks

The rapidly advancing field of Proactive HRC is poised to revolutionize various manufacturing tasks, particularly those related to dis/assembly, drilling, and welding. This Proactive HRC system is specifically designed to enhance task execution between humans and industrial robots in complex manufacturing environments characterized by challenges such as high noise levels, oily workspaces, and constrained spaces. Ensuring the safe and efficient deployment of industrial robotic systems for collaboration in these manufacturing tasks is of paramount importance, with the ultimate goal of boosting productivity and optimizing HRC.

Recent years have witnessed continuous evolution in the field of LLM, opening up new avenues to enhance the planning and interaction capabilities with human-in-the-loop control. Industrial robots, with their high payload capacities, advanced mechanisms, and specialized functions, are well-suited to address various manufacturing challenges. LLM-based Proactive HRC offers the potential to enhance task and motion planning by leveraging extensive training data and deep learning techniques in intricate manufacturing environments. By learning from past experiences, LLM-equipped industrial robots can provide more accurate and efficient decision support, enabling them to collaborate effectively with human counterparts.

Furthermore, LLM enhances their natural language processing and generation capabilities, facilitating more natural and seamless dialog and interaction. This improvement holds the promise of enhancing communication and collaboration between humans and industrial robots, allowing robots to better understand and respond to human instructions and needs, thus optimizing the manufacturing process.. In this envisioned future scenario, individuals without prior experience in robot operation may guide industrial robots to perform tasks through natural language or other forms of interaction, thereby increasing the accessibility and utility of industrial robots across a wider range of users.

11. Evaluation index for Proactive HRC performance

While HRC applications in manufacturing are appearing, a universally accepted evaluation framework for assessing system performance remains absent. The direct transplantation of indicators from automated systems is unjust, given that HRC systems are marked by manual human involvement, which, while critical, can introduce considerable time constraints. Hence, the imperative task is the development of both quantitative evaluation measures and qualitative analytical tools to effectively gauge the performance of HRC systems in the context of flexible automation.

In the evaluation of robot performance, prior quantitative experiments continue remain pertinent, with metrics such as robot execution time, trajectory length, and movement precision. However, for HRC systems, which are distinguished by their human-centric needs and holistic operational dynamics, a broader array of considerations comes into play. Within a mutual-cognitive perspective in Proactive HRC, the formulation of cognitive ergonomics metrics is instrumental. These metrics should assess the physical load borne by humans and the cognitive efforts invested. Evaluating the entire system includes safety mechanisms, like human–robot distance, and the fluency of task execution, especially in the face of dynamic scene changes and uncertain human motions.

Predictability in Proactive HRC can be evaluated through comparative experiments gauging the accuracy of human trajectory prediction and the time horizon for human action prediction. The performance of a predictable HRC process can be corroborated by assessing task plan complexity and waiting times.

The self-organizing intelligence in Proactive HRC introduces additional evaluation methods. Human subject ratings based on questionnaire scores can serve as a quantitative assessment of teamwork involving human operators and robots. Meanwhile, the probability of successful task planning and the transferability of knowledge for new co-working configurations can be introduced as metrics to analyze the performance of self-organizing systems. Besides, there exist numerous other assessment approaches, such as gauging robot end-effector position errors and orientation errors, which provide solutions to test HRC system performance.

Nomenclature

AAD	Assembly Action Dataset
AI	Artificial Intelligence
AGI	Artificial General Intelligence
AGV	Automated Guided Vehicle
ANN	Artificial Neural Network
AR	Augmented Reality
BC	Behavioral Cloning
BiLSTM	Bi-directional Long Short Term Memory
BMS	Battery Management System
BN	Batch Normalization
CAD	Computer-Aided Design
CCA	Canonical Components Analysis
CCD	Charge-Coupled Device
CI	Collaborative Intelligence
CMM	Collaborative Mobile Manipulator
CMNN	Cascade of Multiple Neural Network
CNN	Convolution Neural Network
Cobot	Collaborative robot
CPS	Cyber-Physical System
Dagger	Dataset Aggregation
DDPG	Deep Deterministic Policy Gradient
DMP	Dynamical Movement Primitive
DTW	Dynamic Time Warping
DL	Deep Learning
DoF	Degrees of Freedom
DRL	Deep Reinforcement Learning
DT	Digital Twin
EE	End-Effector
EEG	Electroencephalogram
EM	Expectation Maximization
EMG	Electromyography
EV	Electric Vehicle
EVB	Electric Vehicle Battery
FC	Fully Connected
F/T	Force-Torque
FPN	Feature Pyramid Network
FPS	Frames Per Second
GA	Genetic Algorithm
GAN	Generative Adversarial Network
GCN	Graph Convolutional Network
GMM	Gaussian Mixture Model
GMR	Gaussian Mixture Regression
GPU	Graphics Processing Unit
GRU	Gated Recurrent Unit

HCSM	Human-Centric Smart Manufacturing
HDT	Human Digital Twin
HRC	Human–Robot Collaboration
HRE	Human–Robot–Environment
HRI	Human–Robot Interaction
HRTWE	Human–Robot–Task–Workpiece–Environment
HS	High constant Stiffness
ICP	Iterative Closest Point
ID3	Iterative Dichotomiser 3
IIoT	Industrial Internet of Thing
IMU	Inertial Measurement Unit
IoT	Internet of Thing
IoU	Intersection over Union
IRL	Integral Reinforcement Learning
KG	Knowledge Graph
LDA	Linear Discriminant Analysis
LfD	Learning from Demonstration
LiDAR	Light Detection and Ranging
LLM	Large Language Model
LQR	Linear Quadratic Regulation
LS	Low constant Stiffness
MANO	hand Model with Articulated and Nonrigid deformation
mAP	mean Average Precision
MDP	Markov Decision Process
MMD	Maximum Mean Discrepancy
MR	Mixed Reality
MRR	Mean Reciprocal Ranking
LSTM	Long Short Term Memory
OBB	Oriented Bounding Box
OS	Optimized Stiffness
PCB	Printed Circuit Board
pHRI	physical Human–Robot Interaction
PMC	Probabilistic Model of Compatibility
PSO	Particle Swarm Optimization
QP	Quadratic Program
REBA	Rapid Entire Body Assessment
ReLU	Rectified Linear Unit
RFID	Radio Frequency Identification
RL	Reinforcement Learning
RNN	Recurrent Neural Network
RULA	Rapid Upper Limb Assessment
ROS	Robot Operating System
RRT	Rapidly-exploring Random Tree
SAPNet	Spatial Attention Pyramid Network
SG	Scene Graph
SGD	Stochastic Gradient Descent
SLAM	Simultaneous Localization and Mapping
SME	Small- and Medium-sized Enterprise

SPD	Semipositive Definite
SVR	Support Vector Regression
ST-GCN	Spatio-Temporal Graph Convolutional Network
TSD	Task Sequence Dataset
TSGA	Temporal SubGraph Attention
VIC	Variable Impedance Control
VLN	Vision Language Navigation
VPS	Videos Per Second
VQA	Visual Question Answering
VR	Virtual Reality

Index

Printed and bound by CPI Group (UK) Ltd, Croydon, CR0 4YY

03/10/2024

01040847-0005